WITHDRAWN

St. Louis Community
College

WHAT EVERY ENGINEER SHOULD KNOW ABOUT

ENGINEERING INFORMATION RESOURCES

WHAT EVERY ENGINEER SHOULD KNOW

A Series

Editor

William H. Middendorf

Department of Electrical and Computer Engineering
University of Cincinnati
Cincinnati, Ohio

Vol. 1 What Every Engineer Should Know About Patents, *William G. Konold, Bruce Tittel, Donald F. Frei, and David S. Stallard*

Vol. 2 What Every Engineer Should Know About Product Liability, *James F. Thorpe and William H. Middendorf*

Vol. 3 What Every Engineer Should Know About Microcomputers: Hardware/Software Design: A Step-by-Step Example, *William S. Bennett and Carl F. Evert, Jr.*

Vol. 4 What Every Engineer Should Know About Economic Decision Analysis, *Dean S. Shupe*

Vol. 5 What Every Engineer Should Know About Human Resources Management, *Desmond D. Martin and Richard L. Shell*

Vol. 6 What Every Engineer Should Know About Manufacturing Cost Estimating, *Eric M. Malstrom*

Vol. 7 What Every Engineer Should Know About Inventing, *William H. Middendorf*

Vol. 8 What Every Engineer Should Know About Technology Transfer and Innovation, *Louis N. Mogavero and Robert S. Shane*

Vol. 9 What Every Engineer Should Know About Project Management, *Arnold M. Ruskin and W. Eugene Estes*

Vol. 10 What Every Engineer Should Know About Computer-Aided Design and Computer-Aided Manufacturing, *John K. Krouse*

Vol. 11 What Every Engineer Should Know About Robots, *Maurice I. Zeldman*

Vol. 12 What Every Engineer Should Know About Microcomputer Systems Design and Debugging, *Bill Wray and Bill Crawford*

Vol. 13 What Every Engineer Should Know About Engineering Information Resources, *Margaret T. Schenk and James K. Webster*

Other volumes in preparation

WHAT EVERY ENGINEER SHOULD KNOW ABOUT

ENGINEERING INFORMATION RESOURCES

Margaret T. Schenk
James K. Webster

Science and Engineering Library
State University of New York
Buffalo, New York

MARCEL DEKKER, INC. New York and Basel

Library of Congress Cataloging in Publication Data

Schenk, Margaret T.,
What every engineer should know about engineering information resources.

(What every engineer should know ; v. 13)
Includes bibliographies and index.
1. Engineering–Information services. 2. Technical literature. I. Webster, James K. II. Title. III. Series.
T10.7.S34 1984 620'.0072 84-11350
ISBN 0-8247-7244-X

MARCEL DEKKER, INC.
270 Madison Avenue, New York, New York 10016

Current printing (last digit):
10 9 8 7 6 5 4 3 2 1

PRINTED IN THE UNITED STATES OF AMERICA

Preface

This volume in the *What Every Engineer Should Know* series developed out of a conviction that every engineer should know about engineering information resources, just as he should know about product liability or project management or microcomputers.

No engineer can afford to be unaware of the various types of information and data used in the practice of engineering–handbooks, tables, standards and specifications, technical reports–or to waste expensive time struggling to locate them when the need arises. Existing information represents an investment of manpower and dollars, and duplicating the research that produced it is an expenditure no company or other agency can afford. Francis Bello, the Science Editor of *Fortune* for several years, expressed it well in the September 1960 issue:

"Mankind is learning things so fast that it's a problem how to store information so it can be found when needed. Not finding it costs the U.S. over $1 billion a year."

Technology is advancing at such a rapid rate that engineers must know how to use the literature to keep abreast of developments in their fields. The "life expectancy" of an engineering degree is estimated to be seven years. That degree must be updated with continuing education and monitoring of literature in the field. A change in assignment or in job may necessitate acquiring background knowledge in a new field of interest. Knowing how to locate the information that will supply an overview will save the engineer's company time and money. In addition, the effective use of engineering information resources can lower the cost of research and the development of new products and improvement of existing products. Conversely, the lack of knowledge of what has been done and what is being done in the field can seriously affect an engineer's value to his employer.

Engineers work in a variety of environments. They may be in a small company remote from a large research library, or in a company with a highly specialized information center. If they do not know how to utilize information resources, they are doubly handicapped: They do not know what exists, and they do not know where to go to access the information they need.

Unfortunately, an introduction to engineering literature is not a required course in engineering schools. At best, engineers' knowledge of it, as they leave the college or university, is uneven. To quote from an ad for *Engineering Index*, a long-established and respected literature-indexing service:

"In some schools, engineering students rarely use their library's resources until their senior year, if then. A pity. For learning to use those resources is the student's first step in knowing where to turn for the right answers. Once a practicing engineer, this ability will be vital."

Since most information is in libraries, learning to use library services is unavoidable. *Forbes*, the business magazine, in its February 16, 1981 issue calls libraries "the greatest source of untapped unfathomable riches." Using the sources described in the chapters of this book will help engineers to tap them.

Engineering literature includes a variety of formats—technical reports, periodicals, tables, etc.—all of them essential to engineers. The material in this book is arranged by these formats. Each chapter includes a brief description of the format, and includes a list, not comprehensive but selective, of representative reference tools. We have tried to balance titles that will remain current with those that have several editions, and titles that will serve the current awareness function.

The index is a quick means of locating an item at the moment it is needed. Since the index provides a subject approach, we have chosen to go with the format approach, surveying the various types with which members of the engineering community must be familiar. Further references can be obtained at scientific and technical libraries and from the indexes of *Scientific and Technical Books and Serials in Print*, which is published annually.

Margaret T. Schenk
James K. Webster

Contents

Preface iii

1. Periodicals 1
2. Abstracts and Indexes (Printed and Computerized) 8
3. Non-Bibliographic Databases 19
4. Book and Publisher Information 24
5. Technical Reports 30
6. Handbooks and Manuals 38
7. Tables 53
8. Encyclopedias 60
9. Dictionaries 66
10. Directories (People, Organizations and Products) 73
11. Standards and Specifications 83
12. Patents 100
13. Reviews and Yearbooks 116

14. Dissertations and Theses 122
15. Trade Literature 125
16. Translations 131
17. Conferences 140
18. Professional Societies 146
19. Maps and Atlases 151
20. Statistical Information Sources 157
21. Audio-Visual Materials 161
22. Engineering Software 164
23. License Review 171
24. Preparation of Technical Reports 177
25. Libraries, Information Centers, and Information Brokers 183
Appendix (Publishers' Addresses) 195

Index 205
About the Authors 215

WHAT EVERY ENGINEER SHOULD KNOW ABOUT

ENGINEERING INFORMATION RESOURCES

1

Periodicals

By definition, periodicals are publications with regular frequency, intended to be ongoing or continuous. They may be published weekly, daily, monthly, annually, or at some other periodic interval. They may be scholarly journals, newsletters, government publications, house organs (or company magazines), transactions, or yearbooks. They may be called journals, magazines or periodicals.

They are vital to engineers, because they publish the results of research, within months, enabling engineers to keep current with new developments in materials and processes. In contrast to books, the information in periodicals is briefer, more specific in content, and more current. Periodicals range from newsletters to journals of a highly technical nature.

Periodicals were being published before the United States existed. The Royal Society's *Philosophical Transactions*, still being published today, began publication in 1665. From a few journals in the 17th century, the number of scientific and technical journals being published worldwide has risen to an estimated 60,000. New journals are published as new fields or aspects of engineering evolve. Journals cease publication from time to time for a variety of reasons. New developments may make the focal subject of a journal obsolete. Publishing costs or lack of leadership may cause journals to merge with others or be discontinued.

Periodicals are produced by associations, academic and other institutions, government agencies, industrial firms and other companies, and commercial publishers. Refereed journals are those whose articles have been submitted for review to authorities in the field before they are selected for publication. This evaluative process ensures their accuracy and significance as a contribution to the literature.

Although some are free, most periodicals have an annual or biennial subscription fee payable in advance. Membership in a professional society often entitles the member to a subscription to one or more of the society's publications. Each of the largest engineering societies has one or more periodicals. Examples are *IE*, or *Industrial Engineering*, published by the Institute of Industrial Engineers; a whole series, including *Journal of Applied Mechanics, Journal of Engineering for Industry,* and several others, published by the American Society of Mechanical Engineers, as their *Transactions; AIChE Journal* and the recently created *Energy Progress* and *Plant/Operations Progress*, published by the American Institute of Chemical Engineers; the *Journals* of the various Divisions of the American Society of Civil Engineers. The publications of a society are listed in its entry in the *Encyclopedia of Associations*, to be discussed in a later chapter. Journals, by title, are described in *Ulrich's International Periodicals Directory*, described below.

Some journals are sent free to persons working in appropriate fields. *Materials Handling Engineering* is a case in point. The value to advertisers justifies the cost of publishing and distribution. House organs, the publications of companies for their customers (external organs) or employees (internal organs), are often disseminated to the engineering community as a form of advertising or recruiting.

The chances are that most engineers will not be satisfied with the random selection being sent to them. For awareness of the publishing in their field, they can turn to a list of current journals, issued annually, *Ulrich's International Periodicals Directory*. This directory, which has been published by the R.R. Bowker Company since 1932, has journal titles listed under approximately 200 categories such as engineering (subdivided by type, i.e., electrical, civil, etc), physics, mathematics, etc. Entries for each journal provide information on frequency, publisher, editor(s), price, indexing services that cover articles in the journal, and special features of the periodical, such as book reviews. It includes an index of journals by title and a list of cessations (journals no longer being published). It is updated between editions by *Ulrich's Quarterly*.

Another source of information on current periodicals is the *British Union Catalogue of Periodicals: New Periodical Titles*. It updates the *World List of Scientific Periodicals, Published in the Years 1900–1960*, and gives the holdings of selected British libraries.

Subscriptions to journals may be placed directly with the publisher or handled by a subscription agency. Libraries can provide the catalogs or information that will help in locating these agencies. Libraries will usually fill most of the engineer's information needs, since no engineer can afford to buy all of the books or journals in fields of interest.

Although the libraries that they use will have different methods of informing their users of the periodical titles in the collection, it is well for engineers to be aware of the types of reference tools that are generally available. Most libraries have a list of their holdings in a notebook or a card file or in the library's card catalog. If the library is a unit of a system, there will be a combined record (called a union list) of the holdings of all of the units, and where they are located in the system. The Engineering Societies Library in New York has a major portion of the materials indexed and abstracted in the *Engineering Index*. *CASSI*, the acronym for *Chemical Abstracts Service Source Index*, is an extensive list of the materials indexed and abstracted in *Chemical Abstracts* and libraries holding them. The Engineering Societies Library and Chemical Abstracts Service supply copies of their materials directly to requestors for a fee.

Most engineers have the privilege of requesting interlibrary loan through the libraries they use. The procedure is described in the chapter on libraries and technical information centers. Through interlibrary loan, they can often obtain free or inexpensive copies of journal articles, etc., but when time is at a premium, the immediate accessibility such services as Chemical Abstracts Service offers cannot be underestimated. Copying must comply with the Copyright Act. If the engineer is making the copy, one copy for utilization in research is construed as "fair use" and is permissible.

In 1965 the H. W. Wilson Company published the third edition of the *Union List of Serials in Libraries of the United States and Canada*. It is kept current by *New Serial Titles*, now issued monthly and cumulated four times a year and every five years. This compilation of serials (periodicals plus other continuing publications) makes it possible to determine the closest library holding a periodical and saves time for the library or engineer trying to obtain it. Entries give inception date of the serial and its publisher, title changes, and date of cessation, if appropriate. It lists the libraries in the United States and Canada currently holding volumes of the title. Monthly issues list new periodicals, title changes and ceased titles. Title changes are not unusual. *Journal of Fluid Mechanics* has borne the same title since it began publication in 1956. On the other hand, *Factory Management* has been known as *Factory* and *Modern Manufacturing* during its years of publication.

When an engineer needs an old issue or volume, he can send his request to any of a number of firms handling back issues or back runs. The library can help in supplying names and addresses of these firms.

Most journals have an annual index. Some have cumulative indexes covering from five to fifty years. There is nothing regular about the cumulations or the intervals. There is a cumulated index of publications of the American Institute of Chemical Engineers, *AIChE Publications: Combined Cumulated Index, Subject and Author, 1955–72.* The journal, *Human Factors*, has a cumulated index for volumes 1–22, 1958–1980. The Institute of Electrical and Electronics Engineers (IEEE) publishes an annual index to all of its publications each year, *Index to IEEE Publications*.

Subscribers to journals may frequently receive a bonus in the form of a special issue, a collection of articles on a topic of current interest, a directory of products or manufacturers, or a membership directory. Some journals have special annual issues. An example is *Machine Design's* issue on drives. Most special issues come as part of the subscription, but occasionally supplements must be purchased separately.

There are now a number of specialized journals that are cover-to-cover translations of the original language journals, notably Russian. (Occasionally, an article is lacking, but they are usually inclusive.) The translations can be identified by consulting *Guide to Scientific and Technical Journals in Translation* which lists translated titles cross-referenced to original titles. There are journals in one language; others contain articles in one language with abstracts in another; and still others, a random distribution of articles in two languages. Language should be verified before a subscription is placed by the engineer who wishes to read only English on a regular basis. Language of the text can be checked in *Ulrich's*.

Journals are still serving their original purpose of communication among members of the scientific and technical community. Some day microform may replace today's printed copies. It is easy to mail, easy to read and easy to file. Computer-aided methods of printing will cut down publication time and costs. The future of periodical literature is the "electronic journal," and a few journals, especially the newsletter type, are already available for reading using computer searching services and can be accessed by those with computer terminals.

REFERENCES

British Union Catalogue of Periodicals: New Periodical Titles. Butterworths, London. Quarterly with annual cumulation. Published since 1964.

Updates *World List . . .* (see below). Records new periodical titles since 1960.

Chemical Abstracts Service Source Index (CASSI) 1907–1979 Cumulative (Vols. 1–2).

Updated by quarterly supplements with annual cumulation.

Guide to Scientific and Technical Journals in Translation (2nd Ed.). Carl J. Himmelsbach and Grace E. Brociner (comps.), Special Libraries Association, New York, 1972, 49 pp.

New Serial Titles: A Union List of Serials Commencing Publication After December 31, 1949. Library of Congress, Washington, D.C. Monthly.

Ulrich's International Periodicals Directory: A Classified Guide to Current Periodicals, Foreign and Domestic. R.R. Bowker Co., New York, N.Y. Annual. Published since 1932.

Updated between editions by *Ulrich's Quarterly: A Supplement to Ulrich's International Periodicals Directory and Irregular Serials and Annuals*. Follows format of *Directory and Irregular Serials and Annuals.* Provides information on new serials, title changes and cessations.

World List of Scientific Periodicals Published in the Years 1900–1960 (Vols. 1–3). Peter Brown and G. B. Stratton (eds.), Butterworths, Washington, D.C., 1963–1965.

Updated by *British Union Catalogue: New Periodical Titles* (see above).

2

Abstracts and Indexes (Printed and Computerized)

There are thousands of technical and scientific journals publishing articles every year. No engineer could read them all or even scan them, assuming they were readily attainable. Furthermore, every article is not of importance or of interest to every engineer. That is why abstracts and indexes are indispensable to engineers. Engineers can use them to locate a specific article, find reports and other information on particular subjects or by particular authors, and maintain currency in their special fields of interest by reviewing them regularly. Since there are so many periodicals covering so many subjects, the chances of zeroing in on the right periodical or the right issue that contains the right article or appropriate information is almost nil. Helping to locate articles, etc. by subject or author is the function of indexing and abstract services. This function is analogous to that of the card catalog in locating books.

Of course, not all periodicals are abstracted or even indexed. The percentage is estimated to be 25% to 35% of the published periodicals, but it seems safe to say that this proportion includes most of the major journals. The existence of approximately 2000 abstracting or indexing services is evidence of their value to scientists and engineers. At least one has been developed for each of the major scientific/technical fields. An index service lists the subjects covered in a group of periodicals, supplying the information on title, volume, etc. needed to retrieve the articles. An abstracting service adds summaries of the articles being indexed.

Finding the right indexes or abstracts for locating articles appropriate to a field of interest is the engineer's first step. *Engineering Index* is the most frequently used by engineers. Physicists have *Physics Abstracts*. (That is not to say that electrical engineers, for example, would not find this useful.) While *Chemical Abstracts* is most heavily used by chemists and chemical engineers, this abstract service covers so many fields other than chemistry that it is widely used by all of the sciences and engineering. Some services are highly specialized, as for example, *Ergonomics Abstracts, Zinc Abstracts* or *Abstract Journal in Earthquake Engineering.*

Most abstracting and indexing services are issued at least once a month, some weekly or bi-weekly, so that their users can scan the appropriate sections or indexes to identify articles, reports, etc. to stay current in their subject areas. Most have a similar format. They are usually arranged by subject, but may vary from one service to another, with different subject headings used for the same subject area. Some are arranged by fairly broad subject areas with subdivisions like *Engineering Index*; some by more specific headings like *Applied Science & Technology Index*. Some have their own classification systems like *Electrical and Electronics Abstracts. Current Contents* has a keyword index. Although this service is somewhat different from the others, it is probably most current. There are seven series published concurrently. The series most relevant to engineering, obviously, is *CC/ET&AS* (Engineering, Technology & Applied Sciences), but *CC/PC&ES* (Physical, Chemical & Earth Sciences) and *CC/AB&ES* (Agriculture, Biology & Environmental Sciences) also cover subjects of interest to engineers. *Current Contents* reproduces tables of contents of a broad spectrum of journals. Issues have a directory with full mailing addresses for the senior author of almost every article in the issue. The publisher of *Current Contents*, Institute for Scientific Information, provides, for a fee, the full text of articles listed in the tables of contents.

Abstracting and indexing services include a list of the journals they index. The list spells out the full name of the journal which is usually abbreviated in the entries. Often publisher information is included, or "coden," a system of acronyms for journal titles. It is important that the user consult the list in searching for an article in a particular journal. If that journal is not on the list, it is futile to go further in that index. The next step is to determine what indexing or abstracting services cover the particular title. In *Ulrich's International Periodicals Directory*, described in Chapter 1, the entry for each journal includes the names of the abstracting or indexing services which cover that journal. Some services index periodicals completely, covering letters to the editor, obituaries, brief communications, etc.; others, only the significant articles in an issue.

Entries in the issues of abstracting and indexing services generally include author(s), title of the article (or paper or report), title of the journal (or book or conference proceedings or thesis, etc.), bibliographic data (i.e., volume and issue number, date and pages; report numbers, etc.). Most services have author indexes, and *Engineering Index*, like *Current Contents*, supplies a list of author affiliations. *Chemical Abstracts* has special indexes (ring, formula), appropriate to a chemical index. Individual issues usually contain indexes for the particular issue and cumulate in an annual index by author and subject. Some services have semiannual indexes; some, like *Chemical Abstracts* have a five-year cumulative (collective) index.

Introductory pages in the individual issues describe the scope of coverage, provide instructions for use, and most often, sample entries. Since indexing or abstracting services have unique features, these pages should be read carefully to insure most effective use of the service.

An abstracting service has the added feature of a summary of the information in the article or other item. Some abstracts describe the article briefly; they are known as "indicative" abstracts. Others provide more detail on materials and processes and are known as "informative." The abstract should make it possible for the engineer to determine whether or not to pursue the complete article. Often the word "index" is used in a title while the publication is really an abstract service. *Engineering Index* is an example. Indexing services may appear more quickly, since they do not wait for abstracts to be written.

Printed abstracts and indexes have been in existence for many years, but the computer-stored databases many of which are their counterparts are a comparatively recent development. Currently, the computer-readable database is created first, then used to generate a variety of products and services including printed abstracts and indexes. These databases enable users to conduct searches of the literature, using the computer, with more ease and speed than is possible using printed indexes.

Computer-searching or online systems operate in the same manner as other information retrieval systems such as library card catalogs and printed indexing services. Books, journal articles, reports, etc. are scrutinized and information elements selected from them to create records for the card catalog, the index or the database. These records serve as surrogates or substitutes representing the books, etc. themselves. The engineer retrieves information about the documents, whether they be in the card catalog or the database, then retrieves the actual book, periodical articles, etc. from the library or other document supplier.

There are several considerations which make it wise to evaluate the type of search to be done beforehand. Most computer-searching services available in libraries retrieve literature citations, not actual data. They are the product of vendors such as DIALOG, BRS (Bibliographic Retrieval Service, Inc.) or SDC (Systems Development Corporation). So far, government and industry have created most of the sources of actual data. Some have become available through commercial services, and their number will undoubtedly increase.

Ease of access is an important consideration. In the printed version users must find the proper subject heading that covers the search topic. With the computer, it is possible to use the searcher's terms and search those terms not only via subject headings, but in the titles and the text of the abstracts. Highly intricate searches may be designed, employing several concepts and the complex, logical relation among them. Keywords representing the subject of the search are chosen, then combined in logical sets. The computer responds with citations containing the search terms. Since engineers are paid relatively high salaries, the time saved in locating the appropriate headings and scanning abstracts effects a substantial savings. Computer-searching is not inexpensive, but since "Time is money," engineers may opt for the machine-searching when it is available. However, questions should be carefully formulated and the search strategy carefully devised to retrieve the most relevant information.

Specific topics produce the best search results, i.e., three to fifty citations. (Too many citations are often as frustrating as too few.) Sometimes there are no viable means of obtaining references quickly or sufficient funds to obtain large numbers of citations. With 1200 baud (or faster) terminals, it is feasible to print citations online (or at the computer during the search), so that the engineer has them immediately, but if the number of citations retrieved is too large, printing offline (overnight, by the searching service) is the economical alternative.

Search results can be limited online, as the searcher directs, by dates, by language or by information source. That is, only citations for specific years will be retrieved, only citations in English (or other specified language), only articles from a particular journal or source. Computer-searching can eliminate unrelated aspects, difficult to do in manual searching. (On the other hand, irrelevant citations may appear even in the most carefully designed search. A human searcher is required to read and interpret abstracts and screen out irrelevant material.)

Computer-searching services offer a feature to eliminate repeating the search: Searches already done can be saved and updated regularly by storing the strategy for a small monthly fee.

It may be more cost-effective for librarians or information specialists to perform searches. They can stay current with changes in databases and search techniques and can integrate the language of engineering into the search strategy. However, engineers will be using computers increasingly in design, etc. and may ultimately become end users in the bibliographic databases as well. It is the librarian or information specialist who is most familiar with the thesauri and controlled vocabulary available for many of the databases. (Some thesauri are online; others, printed volumes.)

Not every abstracting or indexing service has a database counterpart, and conversely, not every database has a printed version. Database coverage is diverse and growing daily. Very few databases match their printed counterparts in the years for which they are available. Searches of the printed indexes are required for the years before 1970 for most abstracting and indexing services. The list of current abstracting and indexing services in this chapter shows comparative availability.

There is a long list of indexes and abstracts covering the scientific and technical disciplines. Some of the major ones of special interest to engineers are listed below. If there is a database equivalent, it is given with the earliest data of entries online. Some services that index reports, patents, dissertations and other formats are covered in the chapters describing these formats. The list is not comprehensive. *Ulrich's*, as mentioned above, lists others by title and subjects.

EXAMPLES OF MAJOR CURRENT ABSTRACTS AND INDEXES

Applied Mechanics Reviews. American Society of Mechanical Engineers, New York. Monthly. Published since 1948.

Annual cumulated author and subject tabulation indexes. Classified subject arrangement. Covers periodicals, books, etc. in mechanical engineering and related fields.

Applied Science & Technology Index. H.W. Wilson Co., New York. Monthly (except July). Published since 1922.

Quarterly and annual cumulations. Alphabetical subject listing. The basic, general, least technical guide to English language periodicals in the scientific and technical fields published in the United States, Great Britain and Canada. Many production-oriented journals included.

Chemical Abstracts. American Chemical Society/Chemical Abstracts Service, Columbus, Ohio. Weekly. Published since 1907.

Weekly issues include keyword, patent and author indexes. Annual and collective indexes include author, general subject, chemical substance, formula, ring systems and patent indexes. (Collective indexes are currently issued every five years.) The *Index Guide* and its supplements, for use with the chemical substance and general substance indexes, gives cross-references, synonyms, indexing policy and diagrams, as well as the organization and use of the indexes. It began in Vol. 69. *Chemical Abstracts* is the most comprehensive worldwide abstracting service for chemistry, chemical engineering and related fields, e.g., physics. Computer-searchable from 1967 to the present.

Computer and Control Abstracts. Institution of Electrical Engineers, London. Monthly. Published since 1969.

> *Science Abstracts, Series C.* Author and subject indexes in each issue; semiannually cumulated author and subject indexes. Classified subject arrangement. The major service for computer and control engineering subjects. Computer-searchable, with *Physics Abstracts* and *Electrical and Electronics Abstracts* as INSPEC, from 1969 to the present.

Current Contents: Engineering, Technology & Applied Sciences. Institute for Scientific Information, Philadelphia. Weekly. Published since 1970.

> Tables of contents of selected journals in science and technology. Keyword index provides subject access to the most current material available. Publisher provides full text of articles on request for a fee.

Current Papers in Electrical & Electronic Engineering; Current Papers on Computers and Control. Institution of Electrical Engineers, London. Monthly. Published since 1969.

> These current-awareness journals provide titles and source of articles with the same subject coverage as the corresponding abstract journals (*Science Abstracts. A,B,C*) but without indexes.

Electrical and Electronics Abstracts. Institution of Electrical Engineers, London. Monthly. Published since 1898.

> *Science Abstracts, Series B.* Author and subject indexes cumulated semiannually. Classified subject arrangement. The basic abstracting service for electrical engineering, including computer applications, electronics, current technology, etc. Computer-searchable, with *Physics Abstracts* and *Computer and Control Abstracts*, as INSPEC, from 1969 to the present.

Energy Research Abstracts. U.S. Department of Energy, Technical Information Center, Oak Ridge, Tenn. Semimonthly. Published since 1975.

Title varies slightly. Issues include corporate and personal author indexes, subject, contract and report number indexes; cumulated indexes on microfiche. Covers reports, articles, conference proceedings, books, patents, theses, etc. originated by the Department and its contractors. Availability of items indexed is shown in report number index. Computer-searchable from 1975 to the present.

Engineering Index. Engineering Index, Inc., New York. Monthly. Published since 1883.

The major guide to the world's engineering literature, covering over 4,500 professional and trade journals, conference proceedings, technical reports and books from approximately 40 countries in more than 25 languages. Numbered abstracts listed by subjects with cross-references directing the user to related subjects. Author index. Current Awareness Profiles is a service alerting customers to reports on new developments in their fields of interest. Computer-searchable from 1970 to the present.

INSPEC SDI. Institution of Electrical Engineers, London. Weekly.

A custom selective-dissemination information service matching the interest "profile" of the engineer with the information added to the INSPEC database each week. Subscribers receive cards weekly with the amount of information they specify: index terms, abstract, etc. TOPICS is a similar service based on a set of standard SDI profiles.

International Aerospace Abstracts. Institute of Aerospace Sciences, New York. Monthly. Published since 1961.

Indexes and abstracts the published literature (articles, conference proceedings, etc.) in the field of aeronautics and space science and technology while STAR (see page 35) covers reports.

Metals Abstracts. American Society for Metals, Metals Park, Ohio/Metals Society, London. Monthly. Published since 1968.

Formed by the merger of *Review of Metal Literature* and *Metallurgical Abstracts.* Author, subject indexes in companion publication, *Metals Abstracts Index.*

Computer-searchable from 1966 to the present (includes *Review of Metal Literature*, 1966–7 and *Alloys Index*, from 1974 to the present).

Physics Abstracts. Institute of Electrical Engineers/American Institute of Physics, London. Monthly. Published since 1898.

Science Abstracts, Series A. Author, subject indexes cumulated semiannually. Classified subject arrangement. The basic literature guide for physics and related fields, e.g., electrical engineering. Computer-searchable, with *Electrical and Electronics Abstracts* and *Computer and Control Abstracts*, as INSPEC, from 1969 to the present.

Pollution Abstracts. Cambridge Scientific Abstracts, Bethesda, Md. Monthly. Published since 1970.

Keytalpha (keyterm alphabetical) and author indexes. Includes government reports and documents as well as journals, etc. Divided into ten main areas including air pollution, waste management, noise, radiation, environmental action (technical and non-technical aspects). Computer-searchable from 1970 to the present.

Science Citation Index. Institute for Scientific Information, Philadelphia. Quarterly. Published since 1961.

> *Citation Index* (includes patents), *Source Index* (includes *Corporate Index, Journal Lists & Guides*), and *Permuterm Subject Index*. Interdisciplinary index, identifying authors and their works—if they have been cited or cite other references. *Citation Index*, arranged by cited author, lists subsequent publications which cite an author's articles, etc. *Source Index* lists titles of citing articles with complete bibliographic information. Includes unpublished items, e.g., personal correspondence, studies, communications, pre-publication papers, etc. as well as published serials and monographs. The main lists are by authors' names which can lead to other works on related subjects. A caveat: Only the first author appears in the *Citation Index*. The *Source Index* must be consulted for joint authors. Computer-searchable from 1974 to the present.

There are many major current abstracts and indexes which engineers will use less frequently: *Biological Abstracts*, for biomedical engineering; *Psychological Abstracts*, for industrial engineering; *Bibliography & Index of Geology*, for engineering geology, are examples.

Some libraries include abstracts and indexes in their card catalogs. Some list them with their periodicals holdings. Once the engineer has identified an article of interest by using an indexing or abstracting service, he can determine its availability by using the library resources described in the chapter on periodicals or ordering from one of the agencies such as Engineering Societies Library or Chemical Abstracts Service also noted in that chapter.

3

Non-Bibliographic Databases

The chapter on abstracts and indexes indicates those that can be searched by computer. The databases that are the counterparts of these indexes generate lists of citations of journal articles, conference papers, reports, etc. and are designated "bibliographic." There are other types of databases that supply data, such as the content or text of articles, or provide referral to special types of information.

The referral types usually refer users to organizations or agencies, scientists or engineers, audiovisual or other non-print material, as sources of further information. These and the bibliographic databases are often called "locator tools."

The numeric databases or databanks provide original survey data or representations of data in graphs or tables. They may represent measurements, e.g., volume (in tons, for example) of production of plastics or oil for specified periods of time. They supply information in numbers while the bibliographic or referral bases supply it in words.

Full text databases contain the complete text of a journal article, a standard or a newspaper article. The entire *Harvard Business Review*, for example, is now online.

Further examples of these types of databases follow.

REFERRAL

Electronic Yellow Pages. Produced by Market Data Retrieval, Inc., available on DIALOG.

Series of directories on business and industry. Described fully in chapter on directories.

Licensable Technology. Produced by and available through Dr. Dvorkovitz & Associates on subscription.

Over 30,000 items of technology from worldwide organizations available for licensing. Records list title of technology, licensor, uses of item, degree of development and patent number if one exists.

*Claims**. Produced by IFI/Plenum Data Co., available through DIALOG.

Group of databases on patents covered fully in Chapter 12.

NUMERIC

Chemical Thermodynamics Data Base. Produced by National Bureau of Standards. Information is accessed from NBS Office

---

*Trademark

of Standard Reference Data, A230 Physics Building, Washington, D.C.

Contains recommended values for more than 15,000 inorganic substances: values for enthalpy, heat capacity, free energy of formation and other properties.

EMIS (Electronic Materials Information Service). Produced by the Institution of Electrical Engineers, available online through General Electric Information Services Co.

Numeric data and other information on properties of materials used in design of solid state and other electronic devices.

*ERGODATA**. Produced by Université Rene Descartes, Laboratoire d'Anthropologie Appliquée, online via ERGODATA.

Statistical analysis of human biometric and ergonomic data based on 30 years of anthropometric measurements of world populations. Includes data on human body measurements and space requirements, visual and auditory acuity, etc.

FEDEX (Federal Energy Data Index). Produced by U.S. Department of Energy, available to contractors of the Department of Energy.

Covers publications containing statistical data on energy resource reserves, production, consumption, prices and supply and demand.

Kirk-Othmer/Online. Produced by John Wiley & Sons and accessible online through BRS and DATA-STAR.

A textual-numeric base, with records including the cited

---

*Trademark

references and data from figures and tables in the printed version of the well-known reference tool.

MIDAS (Metals Information Designations and Specifications). Produced by The American Society for Metals and available online through SDC Information Services.

A textual-numeric base drawing data from journals, reports and handbooks. Supplies designations and specification number, including equivalent numbers from various countries, for ferrous and nonferrous alloys. Lists physical properties, uses, forms and manufacturers.

*ORR System of Construction Cost Management**. Produced by Constech, Inc., available online through Control Data Corporation and General Electric Information Services Company on surcharge basis.

Provides cost information at various levels of construction, derived from surveys. Users can adjust the values to their own locations and projects.

Real-Time Weather. Produced by and available online through WSI Corporation on subscription with minimum fee.

Hourly, daily, weekly, monthly, annual and forecast weather data on 4,000 worldwide locations reported by National Weather Service and other government agencies and the World Meteorological Association. Upper-air, river-stage and severe-weather data are among those available.

---

*Trademark

FULL TEXT

ACS Journals Online. Produced by the American Chemical Society, available on BRS.

> Full text of 18 ACS journals online from 1980 to present, updated every two weeks. Entire articles or a reference can be printed.

Air/Water Pollution Report. Produced by Business Publishers, Inc., available on Newsnet Inc. on payment of minimum fee.

> Full text of the newsletter which covers water and air pollution and environmental laws and regulations.

ADDITIONAL READING

A source of information on all types of databases is the following:

Directory of Online Databases. Cuadra Associates, Inc., Santa Monica, Calif. Quarterly. Published since 1979.

4

Book and Publisher Information

While engineers rely heavily on libraries for information resources, they occasionally wish to purchase books and other materials for personal collections. Chapter 6 notes the desirability of owning appropriate handbooks, for example. Every field of engineering has its classics, and as positions and interests change, it becomes necessary to acquire books to provide background knowledge in the new area. Fortunate indeed is the engineer who lives or works near a large library and a good bookseller. That engineer can borrow from the former and buy or order from the latter. If an engineer must order independently, knowing how to determine the availability of titles, i.e., what is in print or in stock is essential.

Once the engineer has joined a society, bought a book or attended a conference, the engineer's name will probably be added to the mailing list of companies publishing in the science and technology areas. These mailings, plus book reviews and advertisments in current issues of journals, assure a steady flow of news concerning new books, journals and even technical reports, although there are special sources for the last. Publishers' announcements are frequent and some are directed only to special groups such as geologists, chemists or engineers. Publishers' brochures announce new publications. Their catalogs list those in stock and therefore purchasable. They list addresses, toll-free or regular phone numbers, sales policies and order information, and generally include order forms.

Large publishers usually have a customer services department that will handle orders via letter or phone. Many will send books on approval. This may entail the labor of returning unwanted volumes and the cost of postage, but when there is doubt of the value to the engineer of an expensive book it is worth the effort. Publishers' services vary. Standing or continuing orders can be arranged with most publishers, so that books and other publications are mailed to the engineer or library as soon as a new edition or a new volume in a series is published.

There are book clubs which offer news of new books in specific fields and discounts on purchases. Range of materials available and terms vary.

Societies almost invariably offer their members a substantial savings on their publications. Societies are heavy publishers. They, too, have mailings to inform members and others of newly published materials. Their addresses can be found in the *Encyclopedia of Associations.* (See page 147.)

One of the best methods for selecting books is reading reviews. Many technical journals include book reviews, and there are two publications devoted exclusively to reviews of technical and scientific books: *Technical Book Review Index* and *New Technical Books.* Book reviews may also be located by computer search: *Book Review Index* is computer-searchable from 1969 to the present. Unfortunately, book reviews are fairly slow in appearing, but they are most often written by experts and the reader has the advantage of their evaluations before investing.

Acquiring older material calls for the use of reference tools that are in almost every library: *Books in Print, Scientific and Technical Books and Serials in Print, Forthcoming Books.* The publication date is no indication of the availability of a book. *Books in Print* lists the books in English published or distributed in the United States and in print or in stock with full bibliographic or ordering information. A subject guide provides this approach to the in-print books listed in the author and title volumes. In 1983, the set consisted of three author volumes and three title volumes. The *Subject Guide to Books in Print*, as well, had three volumes. *Books in Print* is also computer-searchable.

A spinoff from the set is restricted to scientific and technical titles: *Scientific and Technical Books & Serials in Print* which provides the same access in three volumes. Both reference tools include a list of publishers' addresses and phone numbers. There is even a publication titled *Forthcoming Books* whose issues supplement and update *Books in Print*, supplying information on titles in preparation due to be published within five months. It has the same indexes as *Books in Print* and is published bimonthly.

For retrospective purchases there is *Pure and Applied Science Books 1876–1982.* The date 1876 was chosen, because it was that year in which the Dewey Decimal Classification was published. This set has titles arranged by subject and has an author index.

The volumes of the catalog of the Engineering Societies Library are rather complicated for non-librarians to use, since their arrangement is by the aforementioned Dewey Decimal Classification numbers. However, recent additions to the collection have been published annually since 1974 as *Bibliographic Guide to Technology* and offer an extensive selection of conference proceedings in the Engineering Societies Library and technical books added to the New York Public Library for the engineer to survey. The Engineering Societies Library is open to the public, loans books to members of affiliated societies (including student members) and provides photocopy, with fees attached to both services.

Occasionally, an engineer needs a book which is out of print (*op*), or no longer in stock. Such books can often be located, but they command a price much higher than the original, since locating them requires a search by a dealer specializing in out-of-print books. Librarians can help the engineer identify out-of-print dealers and supply their addresses and phone numbers.

Some dealers specialize in foreign books. Dealers in foreign countries may be contacted directly or through affiliations in the United States. Again, librarians can assist in supplying current addresses and phone numbers. Publications similar to *Books in Print* exist in other countries, e.g., *British Books in Print, Bibliographie de la France. Livres* and *Deutsche Bibliographie.* There is also an *International Books in Print* which lists books in English published outside the United States and the United Kingdom.

A highly selective list of publishers who produce a high percent of scientific and technical titles are listed with their addresses and phone numbers in the appendix of this book, "Publishers' Addresses." The list of publishers in the United States in *Books in Print* should be consulted for a comprehensive list.

The chapters on periodicals, technical reports, standards, patents and maps will supply information on purchasing these forms.

REFERENCES

Bibliographic Guide to Technology. G. K. Hall, London. Annual. Published since 1974.

Continuation of annual supplements to *Classed Subject Catalog of the Engineering Societies Library.*

Bibliographie de la France. Livres . . . Cercle de Librarie, Paris. Monthly. Published since 1972.

Books in Print. R. R. Bowker Co., New York. Annual. Published since 1948.

> A comprehensive list of books in English published or distributed in the U.S. and currently in print or available for purchase. Author and title volumes and companion volumes of *Subject Guide to Books in Print* provide three points of access.

British Books in Print. J. Whitaker & Sons, London. Annual. Published since 1874.

> Author and subject indexes in one alphabet.

Deutsche Bibliographie; Wochentliches–Verzeichnis. Buchhändler-Vereinigung, Frankfurt. Weekly. (Cumulated monthly and quarterly.) Published since 1947.

Forthcoming Books. R. R. Bowker Co., New York. Bimonthly. Published since 1966.

> Updates *Books in Print*, listing books published since the latest edition of *Books in Print*, those due to appear in the next five months and those whose publication has been postponed. Author and title indexes.

New Technical Books: A Selective List with Descriptive Annotations. New York Public Library, New York. 10 issues per year. Published since 1915.

> Lists new books added to the Library. Issues arranged by subject with author and subject indexes.

Pure and Applied Science Books 1876–1982. (Vols. 1–6) R. R. Bowker Co., New York, 1982.

Over 220,000 titles in technology and the physical and biological sciences. Author, title and subject indexes.

Scientific and Technical Books & Serials in Print. R. R. Bowker Co., New York. Annual. Published since 1974.

Author, title and subject indexes to books and title and subject indexes to serials in the sci/tech fields.

Technical Book Review Index. Carnegie Library, Science and Technology Dept., Pittsburgh, Pa. Monthly. Published under the sponsorship of Special Libraries Association since 1935.

Aid to locating reviews in journals. Includes excerpts from reviews.

5

Technical Reports

It is probably unnecessary to explain to practicing engineers what a technical report is, since it is quite likely that they have been using this form of literature more than any other.

For the record, however, here is a good working definition: A technical report is a report of research results prepared by a scientist or engineer that is sent to the organization that sponsored the research. The report is intended for use within the investigator's and sponsor's organizations, and is not refereed or edited to the same standard that a journal article or book is. The technical report is usually printed or duplicated by the investigator's or sponsor's organizations, and is often disseminated to requestors in microfiche format. A technical report may contain details about equipment, procedures, and detailed experimental results that cannot be printed in a journal or book because of space restrictions.

Technical reports, along with papers presented at conferences, provide the engineer with the very latest information on a subject. The general rule of thumb for currency of information is that some research will be described first in a technical report or conference paper, then some time later refined for publication in a periodical, and much later published or included in a book.

Citations to technical reports look quite different from references to periodical articles or books. Among the items usually appearing in a citation that distinguish a technical report are: corporate organization names, sponsoring organization, contract or grant numbers, and various report or accession numbers. The city of publication and the name of the publisher are rarely given.

Here are examples of citations to the three basic types of literature:

Book

Theory and application of digital signal processing. L. R. Rabiner and B. Gold, Prentice-Hall, New York, 1975.

Periodical

Muche, C. E., et al. New techniques applied to air traffic control radars. *IEEE Proc. 62*:716–723, 1974.

Technical Report

Cartledge, L. and O'Donnell, R. M. Description and performance evaluation of the moving target detector. *MIT Lincoln Lab. Project Rept. ATC–69*, Mar. 8, 1977. (FAA/RD–76/190) Contract F19628–76–C–0002. Available from NTIS as AD–A040–055.

Several federal agencies index technical reports, publish indexes and abstract journals to announce them, and make copies of the reports available to requestors.

The foremost of these is the National Technical Information Service (NTIS), which was formerly known as the Clearinghouse for Federal Scientific and Technical Information, and, prior to that, as the Publication Board. NTIS, an agency of the U.S. Department of Commerce, is the central source for the public sale of U.S. Government-sponsored research, development, and engineering reports, as well as foreign technical reports and other analyses prepared by national and local government agencies, their contractors, or grantees. 70,000 new reports of completed research are added to the NTIS data base annually. The NTIS abstract journal, *Government Reports Announcements & Index*, described in more detail below, is the major vehicle for announcing and locating the reports handled by this agency.

It is important for engineers to know that NTIS will accept, announce, and disseminate reports contributed to them, even if they aren't sponsored by the government. Documents resulting from commercial or internal research funding may thus be given international exposure, and the engineer's company is spared the expense of distributing copies to requestors. For further information about submitting reports to NTIS, contact them at the following address:

Acquisitions Manager
5285 Port Royal Road
Springfield, VA 22161
703–487–4785

Listed below are the major report-announcing indexes. In addition to these, a few other indexes whose major focus is periodical literature, such as *Engineering Index* and *Chemical*

Abstracts, include some technical reports. The descriptions here are for the printed versions of these indexes, but all are also available in computer-searchable form, as indicated.

INDEXES TO TECHNICAL REPORTS

Energy Research Abstracts. U.S. Department of Energy, Technical Information Center, Oak Ridge, Tenn. Published twice a month since 1975.

> Issues include corporate and personal indexes, subject, contract and report number indexes. Semiannual and annual cumulated indexes. Covers reports, articles, conference proceedings, books, patents, theses originated by the Department of Energy and its contractors. (Supersedes in part *Nuclear Science Abstracts*).
>
> Available for computer-searching from 1975 through SDC Information Services, Dialog Information Services, and the DOE/RECON System.

Government Reports Announcements & Index. U.S. Department of Commerce, National Technical Information Service, Washington, D.C. Published twice a month since 1946.

> Subject, personal author, corporate author, contract and report number indexes cumulated annually. Principal guide to technical reports produced by government agencies and contractors. Limited and classified Department of Defense reports are announced in the Defense Technical Information Center's *Technical Abstract Bulletin*.
>
> Available for computer-searching from 1964 through SDC Information Services and Dialog Information Services, and from 1969 through BRS.

INIS Atomindex. International Atomic Energy Agency, Vienna. Published twice a month since 1976.

INIS (International Nuclear Information System) is a cooperative decentralized information system of approximately 100 nations. *INIS Atomindex* includes most major international documents on nuclear science. (Supersedes in part *Nuclear Science Abstracts*).

Available for computer-searching through SDC Information Services, Dialog Information Services, and the DOE/RECON System.

Monthly Catalog of United States Government Publications. U.S. Government Printing Office, Superintendent of Documents, Washington, D.C. Published monthly since 1895.

Author, title, subject, series/report indexes cumulated semi-annually. Lists publications of U.S. departments and agencies, including Congressional publications. Entries are by Superintendent of Documents classification number, generally, alphabetical by department. Library of Congress cataloging form has been used since 1976. Annual list of serial publications.

Available for computer-searching from July 1976 through BRS and Dialog Information Services.

Nuclear Science Abstracts. U.S. Atomic Energy Commission/ Energy Research & Development Administration, Oak Ridge, Tenn. Published twice a month from 1946 to 1976.

Personal author, corporate author, subject report number indexes, cumulated annually. Classified subject arrangement. Main guide to AEC/ERDA reports and nuclear science research. (Superseded by *Energy Research Abstracts* and *INIS Atomindex*.)

Available for computer-searching through the DOE/ RECON System.

Scientific and Technical Aerospace Reports. National Aeronautics and Space Administration, Scientific and Technical Information Facility, B.W.I. Airport, Maryland. Published twice a month since 1963.

> Semiannual and annual cumulated indexes: subject, personal author, corporate source, contract and report number. Covers NASA and other government agencies' and contractors' reports, patents, translations, dissertations.
>
> Presently available for computer-searching only by mail request or on-line for NASA contractors. This file is expected to be available from a commercial vendor in 1984.

Selected Water Resources Abstracts. U.S. Department of the Interior, Geological Survey, Reston, Va. Published monthly since 1968.

> Personal author, organization, subject, cumulated annually. Includes abstracts of current and earlier pertinent monographs, journal articles, reports, and other publication formats.
>
> Available for computer-searching through Dialog Information Services and the DOE/RECON System.

Technical Abstract Bulletin. U.S. Department of Defense, Defense Technical Information Center, Alexandria, Va. Published twice a month since 1953.

> Corporate author, subject, title, personal author, contract, report number indexes; cumulated quarterly, semiannually and annually. Announces the availability of technical reports acquired by DTIC. Contains descriptions of classified and unclassified documents that *cannot* be released to the general public. (Department of Defense reports whose distribution is not restricted in

any way, are announced in *Government Reports Announcements & Index*).

This index has recently been designated "Confidential" by DTIC.

Available for computer-searching to Department of Defense registered users, directly from DTIC.

OBTAINING COPIES OF TECHNICAL REPORTS

Once a citation to a technical report has been located in any of the above-mentioned indexes, obtaining a copy is a relatively straightforward process. Each index provides clear information about acquiring reports.

Most reports listed in *Government Reports Announcements and Index* are available for purchase from NTIS, in paper copy or microfiche format. Order forms are included in every issue, and procedures for calling in rush orders and using credit cards are indicated.

It should be noted that many reports that are announced in *Energy Research Abstracts, Scientific and Technical Aerospace Reports, Monthly Catalog of United States Government Publications*, and *Selected Water Resources Abstracts* may also appear in *Government Reports Announcements & Index*, and thus are available from NTIS. In fact, many of the citations in these other indexes will indicate that copies are available from NTIS.

NTIS is gradually becoming the major report-disseminating agency in the United States, but there are still substantial numbers of documents that can only be obtained from the Department of Energy, the Superintendent of Documents, the Defense Technical Information Center, etc. To reiterate, the indexes that these agencies produce not only announce reports, but explain acquisition procedures as well.

ADDITIONAL READING

Role of Technical Reports in Sci-Tech Libraries. *Science & Technology Libraries*, Vol. 1, No. 4, Summer 1981.

A special issue containing six articles on technical reports:

a. Introduction: Role of Technical Reports in Sci-Tech Libraries. Irving M. Klempner (pp. 3–4).

b. Interaction Within the Technical Reports Community. Ruth S. Smith (pp. 5–18).

c. Some Aspects of Technical Report Processing by Federal Agencies. Madeline M. Henderson (pp. 19–26).

d. Managing the Bell Laboratories Technical Report Service. Amy Wang and Diane M. Alimena (pp. 27–40).

e. Three Technical Report Printed Indexes: A Comparative Study. Susan Copeland (pp. 41–53).

f. Managing Exxon's Technical Reports. Karen Landsberg and Ben H. Weil (pp. 55–64).

6

Handbooks and Manuals

The terms handbook, manual and data book are used interchangeably. "Manus," the Latin word for hand, is the origin of the word manual. Manual is often the name given laboratory workbooks, how-to-do-it books, publications accompanying equipment or books an engineer might keep on the desk for frequent reference. The latter are often called desk handbooks. Many books called "encyclopedias" may also be characterized as handbooks. Some characterized as "handbooks" might be more aptly described as subject dictionaries. Often periodicals or journals will publish special issues or contain special sections which qualify as handbooks.

Although their scope is comprehensive, handbooks are compact sources for the significant information on a subject.

They usually concentrate on one subject, i.e., chemical engineering, electronics, steel. Typical handbooks listed in this chapter assemble tables, charts, glossaries, formulas, statistical data, often a history of the subject, in one convenient volume. Most are divided into sections or chapters based on classification of the subject matter in the engineering field, often written by experts in the field. To be maximally useful, they should have comprehensive indexes. Their chief utility is supplying facts and data, and the index facilitates location of these. Most contain bibliographies or references to direct the user to further information.

Engineers have had handbooks for years. They were frequently called *vade mecum* (Latin for "carry with me") or pocket books. The more popular ones have had several editions and are updated frequently by new editions. However, even older editions are useful for their established facts and data. They are usually reasonably priced for the amount of information they contain, so most libraries or company engineering departments can afford the most up-to-date information. The civil, mechanical and chemical engineering fields all have well-known handbooks. The following list includes these. Others can be located in library card catalogs under the subject, with the subheading, "Handbooks, manuals, etc.". There are also a legion of handbooks on more specific subjects such as air conditioning, motors, safety, etc.

The books listed here have been selected as examples, the topics of the handbooks on specific subjects chosen at random. Neither list is intended to be comprehensive or even representative. Many more handbooks and manuals on broad and specific engineering subjects can be located through *Scientific and Technical Books and Serials in Print* which indicates their availability and provides the information necessary for ordering.

Inclusion in the list does not imply the consideration of these handbooks and manuals as more authoritative or useful than similar titles. They have been chosen to serve as examples of various subjects or format.

Several handbooks whose contents are chiefly tables are listed in the chapter on tables.

SELECTED EXAMPLES OF HANDBOOKS FOR THE MAJOR ENGINEERING FIELDS

Chemical Engineers' Handbook (5th Ed.). Robert H. Perry and Cecil H. Chilton (eds.), McGraw-Hill Book Co., New York, 1973. (6th Ed. planned for 1984)

> Material ranges from mathematics and mathematical tables to more specialized sections on psychometry, process machinery, drives and distillation.

Electronics Engineers' Handbook (2nd Ed.). Donald G. Fink and Donald Christiansen (eds.), McGraw-Hill Book Co., New York, 1982.

> Sections on principles, materials, devices, circuits, systems, etc. Greatly expanded over first edition due to increased impact of integrated circuits.

Fundamentals Handbook of Electrical and Computer Engineering (Vols. 1–3). Sheldon L. Chang (ed.), John Wiley & Sons, Inc., New York, 1982–83.

> Volumes treat core areas of electrical and computer engineering with emphasis on system, device and circuit design. Provides quick review of basic mathematics, physics and statistical methods.

Handbook of Industrial Engineering. Gavriel Salvendy (ed.), John Wiley & Sons, Inc., New York, 1982, 1400 pp.

> Comprehensive coverage of performance measurement, manufacturing engineering, quality control, engineering economy.

Mechanical Engineers' Handbook (8th Ed.). Theodore Baumeister (ed.), McGraw-Hill Book Co., New York, 1978.

Commonly known as Marks handbook. Supplies reference data on cryogenics, strength of materials, power sources, mechanisms, etc. Tables and data on all aspects of mechanical engineering.

Standard Handbook for Civil Engineers (3rd Ed.). Frederick S. Merritt (ed.), McGraw-Hill Book Co., New York, 1983.

Information on construction management, materials, structural theory and design. Covers various types of civil engineering, i.e., building, highway, water, etc. Third edition expands treatment of environmental control.

SELECTED EXAMPLES OF HANDBOOKS AND MANUALS ON SPECIFIC SUBJECTS

Accident Prevention

Accident Prevention Manual for Industrial Operations (8th Ed.). National Safety Council, Chicago, 1980, 760 pp.

Everything the safety engineer needs to know: Sections on the Occupational Safety and Health Act of 1970, training, human factors engineering, safety equipment. Includes a directory of service organizations, government agencies, educational institutions and standards/specifications groups.

Aerosols

Handbook of Aerosol Technology (2nd Ed.). Paul A. Sanders, Van Nostrand Reinhold Co., New York, 1979, 526 pp.

Covers homogeneous systems, emulsions, foams, suspensions. Emphasis on alternative fluorocarbons (toxicology and properties), aerosol valves and containers.

Alloys

Engineering Alloys (6th Ed.). Norman E. Woldman, Van Nostrand Reinhold Co., New York, 1979, 1815 pp.

Alloys listed by serial number giving trade names, composition, properties, uses; trade name list with corresponding serial numbers. Index of manufacturers and properties.

Automotive

SAE Handbook. Society of Automotive Engineers, Warrendale, Pa. Annual. Published since 1902.

SAE recommended practices, standards and information reports on metals and nonmetallic materials, electrical equipment, powerplant and other components for vehicle design.

Building

Mechanical and Electrical Equipment for Buildings (6th Ed.). William J. McGuinness et al., John Wiley & Sons, Inc., New York, 1980, 1336 pp.

Sections on heating/cooling, lighting, acoustics, etc. New chapters on energy-climate-site relationships.

Building Construction Cost Data Book. R.S. Means Co., Kingston, Mass. Annual. Published since 1942.

Provides average unit prices on concrete, wood, plastics and other building construction items.

Earthquake Protection of Essential Building Equipment: Design, Engineering Installation. John Wiley & Sons, Inc., New York, 1981, 464 pp.

Detailed description of seismic qualification methods

and categories for equipment, design, specification models; installation details.

Circuits

Designer's Handbook of Integrated Circuits. Arthur B. Williams (ed.), McGraw-Hill Book Co., New York, 1984.

Comprehensive collection of "most popular and useful" IC device selection charts. Arranged by families of ICs.

Encyclopedia of Integrated Circuits: A Practical Handbook of Essential Reference Data. Walter J. Buchsbaum, Prentice-Hall, Inc., Englewood Cliffs, N.J. 1981, 420 pp.

Covers integrated circuits "from analog to interface." Over 1000 entries. List of manufacturers. Glossary.

Special Circuits Ready-Reference. John Markus, McGraw-Hill Book Co., New York, 1982, 234 pp.

Each circuit has type numbers or value of all important components, an identifying title, a brief description, performance data and suggested applications.

Concrete

ACI Manual of Concrete Practice. American Concrete Institute, Detroit. Annual. Published since 1967.

Volumes cover materials and general properties of concrete pavements, concrete in buildings, bridges and masonry.

Controls

Handbook of Controls and Instrumentation. John D. Lenk, Prentice-Hall, Inc., Englewood Cliffs, N.J., 1980, 310 pp.

Covers operating principles of control and instrumentation devices including sensors.

Drives

Mechanical Drives: Reference Issue. Special issue of *Machine Design*, usually published each June.

Articles on drives, bearings, seals.

Electronic Systems

Handbook of Electronic Systems Design. Charles A. Harper, McGraw-Hill Book Co., New York, 1980, 663 pp.

Extensive treatment of electronic systems. Sections on computers, communication, radar, measurement, digital systems.

Energy

Energy Deskbook. Samuel Glasstone, U.S. Department of Energy, Technical Information Center, Oak Ridge, Tenn., 1982, 453 pp.

Material arranged in alphabetical order by subject. Encyclopedic-type articles. No index, but adequate table of contents. (DOE document no. DOE/IR/05114-1)

Environment

CRC Handbook of Environmental Control. (Vols. 1-5) Richard G. Bond and Conrad P. Straub (eds.), CRC Press, Boca Raton, Fla., 1973-74.

Data on pollutants. Social and economic aspects of air pollution, solid wastes, water supply and treatment, and wastewater treatment and disposal.

Formulas

Engineering Formulas (4th Ed.). Kurt Gieck, McGraw-Hill Book Co., New York, 1983.

Brief guide to important technical and mathematical formulas.

Gas Turbines

Gas Turbine Engineering Handbook. Meherwan P. Boyce, Gulf Publishing Co., Houston, 1982, 603 pp.

Design, fabrication, installation, operation and maintenance of gas turbines.

Gases

Handbook of Compressed Gases (2nd Ed.). Compressed Gas Association, Van Nostrand Reinhold Co., New York, 1980, 507 pp.

Basic information on compressed gases: uses, transportation, safety, rules and regulations.

Heat Transfer

Heat Transfer and Fluid Flow Data Books. General Electric Co., Schenectady, N.Y. Looseleaf, updated irregularly. Published since 1970.

Information designed to cover those aspects of heat transfer and fluid flow determined from experience to be used most often by engineers in design and development work.

Heating/Cooling

ASHRAE Handbook and Product Directory (Vols. 1–4). American Society of Heating, Refrigerating and Air Conditioning Engineers, New York. Volumes on *Applications, Equipment, Fundamentals* and *Systems* updated annually on a rotating basis. Published since 1922.

Composite index. Theory, materials, etc. covered.

Highways

Handbook of Highway Engineering. Robert F. Baker (ed.), Van Nostrand Reinhold Co., New York, 1975, 894 pp.

Sections cover urban transportation, planning, geometric design standards and supply photos, diagrams, etc. as well as data.

Lasers

Safety with Lasers and Other Optical Sources: A Comprehensive Handbook. David Sliney and Myron Wohlbarst, Plenum Publishing Corp., New York, 1980, 1035 pp.

Information on hazards to eye and skin and protection of these areas. Laser standards, classifications.

Liability (Product)

Product Safety and Liability. John Kolb and Steven S. Ross, McGraw-Hill Book Co., New York, 1980, 688 pp.

Material on product liability law. Traces product through development, manufacture and marketing phases. List of toxic substances. Appropriate standards.

Mathematics

CRC Handbook of Mathematical Sciences (5th Ed.). William H. Beyer (ed.) CRC Press, Boca Raton, Fla, 1978, 982 pp.

Sections on constants, conversion factors, numerical methods, probability and statistics, financial tables. Symbols and abbreviations.

Handbook of Applied Mathematics: Selected Results and Methods (2nd Ed.). Carl E. Pearson (ed.), Van Nostrand Reinhold Co., New York, 1983, 1307 pp.

Emphasis on analysis and technique. Chapters on formulas, equations, mathematical models, etc.

Mechanisms

Machinery's Handbook: A Reference Book for the Mechanical Engineer, Draftsman, Toolmaker and Machinist (21st Ed.). Erik Oberg and Franklin D. Jones (eds.), Industrial Press, New York, 1979, 2428 pp.

Data selected for inclusion on basis of requirement of design and production engineers in small and large industry.

Metals

Metals Handbook (9th Ed.). American Society for Metals, Metals Park, Ohio, 1978–1983.

Five volumes of 9th ed. issued 1978–83: Extensive information on properties and selection of iron and steel, nonferrous alloys and pure metals; heat treating; surfaces and coating.

Occupational Health

Occupational Health Law: A Guide for Industry. Joseph La Dou (ed.), Marcel Dekker, Inc., New York, 1981, 214 pp.

Material on OSHA and other regulatory agencies; health organization and practices within institutions, etc.; legal aspects.

Microprocessors

Microprocessors Applications Handbook. David F. Stout, McGraw-Hill Book Co., New York, 1982, 477 pp.

Applications contributed by specialists in various fields. Chapters cover both hardware and software aspects of microprocessor system design and specific design information.

Noise

Handbook of Noise Control (2nd Ed.). Cyril M. Harris (ed.). McGraw-Hill Book Co., New York, 1979, 600 pp.

Designed to make technical information comprehensible to engineers, builders, health officials, etc. Covers types of noise, i.e., aircraft, motor vehicle, etc.; noise measurement and regulation.

Seals

Seals and Sealing Handbook. R.H. Warring, Gulf Publishing Corp., Houston, 1981, 431 pp.

General and special methods of sealing: fundamentals, principles, selection and applications.

Soil

Soil Mechanics Technology. Marcus M. Truitt, Prentice-Hall, Inc., Englewood Cliffs, N.J., 1983, 284 pp.

Theoretical concepts and design applications.

Solar Energy

Solar Cell Array Design Handbook: The Principles and Technology of Photovoltaic Energy Conversion. H.S. Rauschenbach, Van Nostrand Reinhold Co., New York, 1980, 549 pp.

Practical information on photovoltaic energy conversion technology. Sections on array analysis, building blocks, etc. Support data including properties.

Solar Energy Handbook. Jan F. Kreider and Frank Kreith (eds.), McGraw-Hill Book Co., New York, 1981, 1100 pp.

Comprehensive treatment of solar conversion systems, thermal collectors, etc. Emphasizes applications and attempts to provide information of a permanent nature.

Steel

Handbook of Comparative World Steel Standards: U.S.A., United Kingdom, West Germany, France, U.S.S.R., Japan. International Technical Information Institute, Tokyo, 1980, 499 pp.

Steel standards and specifications of six major countries. Tables show country, standards and grade.

Manual of Steel Construction (8th Ed.). American Institute of Steel Construction, Chicago, 1980, 924 pp.

Use of fabricated structural steel: dimensions and properties of shapes, plates, bars, etc.; beam and girder design; connections; specifications and codes.

Structures

Structural Engineering Handbook (2nd Ed.). E.H. and C.N. Gaylord (eds.), McGraw-Hill Book Co., New York, 1979, 600 pp.

Sections on planning, design and construction of foundations, concrete, building and various types of structures.

Toxic Substances

Dangerous Properties of Industrial Materials (6th Ed.). N. Irving Sax, Van Nostrand Reinhold Co., New York, 1984, 1118 pp.

Lists hazard information for nearly 19,000 common industrial laboratory materials.

Toxic and Hazardous Industrial Chemical Safety Manual for Handling and Disposal with Toxicity and Hazard Data. International Technical Information Institute, Tokyo, 1981, 646 pp.

Data on treatment and disposal of approximately 700 industrial chemicals. Gives synonyms, uses, properties and other data. Appendix with 1000 additional chemicals.

Valves

Valve Selection Handbook. R.W. Zappe, Gulf Publishing Corp., Houston, 1981, 278 pp.

General information on valves; sections on manual valves, check valves, pressure relief valves.

Very Large-Scale Integration

VLSI Design: Problems and Tools. IEEE Proceedings. Vol. 71, No. 1, January 1983.

This issue of an Institute of Electrical and Electronic Engineers periodical is only one of several special issues dealing with VLSI which have appeared in various journals.

Wastewater

Handbook of Wastewater Treatment Processes. Arnold S. Vernick and Elwood C. Walter (eds.), Marcel Dekker, Inc., New York, 1981, 255 pp.

Fact sheets (two-page summaries) give applications, environmental impact, equipment, chemicals involved in the collection, treatment and disposal of wastewater.

Standard Methods for the Examination of Water and Wastewater (15th Ed.). American Public Health Association, Washington, D.C., 1980, 1200 pp.

Information on proper execution of examination procedures. Sections on determination of metals, inorganic nonmetallic and organic constituents, automated laboratory analysis, biological examination of water, tests for radioactivity and aquatic organisms. Increasing coverage of laboratory safety and quality assurance.

Wear

Wear Control Handbook. M.B. Peterson and W.O. Winer (eds.), American Society of Mechanical Engineers, New York, 1980, 1358 pp.

Wear data for use in machinery design and maintenance. Articles by experts on components and materials.

Wind

Windpower: A Handbook of Wind Energy Conversion Systems. V. Daniel Hunt, Van Nostrand Reinhold Co., New York, 1981, 610 pp.

Sections on history, conversion systems, equipment, commercialization, federal programs. Glossary.

SELECTED EXAMPLES OF HANDBOOKS COVERING SEVERAL FIELDS

Composite Index for CRC Handbooks (2nd Ed.). CRC Press, Boca Raton, Fla., 1977, 1111 pp.

Part 1: Subject entries. Part 2: Chemical substance entries. Both parts refer user to appropriate handbook, volume and pages.

Handbook of Chemistry and Physics. CRC Press, Boca Raton, Fla. Annual. Published since 1913.

Data useful to all scientific and technical fields. Properties of materials; special sections of mathematical tables and sources of critical data.

Kempe's Engineers' Yearbook. Morgan-Grampian, London. Annual. Published since 1894.

Deals with all types of engineering. Covers legal aspects of engineering, including patents, materials, industrial safety, etc.

7

Tables

Tables have been defined as compact displays of data. Conversion tables, gas tables, tables of properties are in handbooks, encyclopedias, and reports, even in journals. (Some of the chemical journals include extra data on microfiche on the back cover of the issue to save space while providing more data.)

The need for evaluated, even critically selected, numerical data was recognized as early as the 1920's. ("Critical," it should be added, connotes the "best" value derived by contributing scientists from current information of the period.) The first response to the need for such data was the *International Critical Tables* (ICT) developed by the National Research Council and published by McGraw-Hill from 1926–1933. The eight volume set is an extensive compilation of physical property data. ICT will not be updated, but since it is important that the scientific and technical community have current data, the National Bureau of Standards established in 1963 the National Standard Reference Data System. Great Britain and the USSR have comparable programs, so it is obvious that the need for critical data is universal. There are organizations or systems that generate or critically evaluate data. The National Standard Reference Data System is an example of such an original source. The NSRDS is the United States' centralized government agency for the collection and distribution of physical data. It accumulates information on the physical and chemical properties of matter and materials basic to nuclear, atomic, molecular and solid-state physics; thermodynamics, transport phenomena; chemical kinetics; mechanical properties. Publications are for sale by the Superintendent of Documents, Washington, D.C.

After critical evaluation for significance and reliability, the data becomes part of the NBS data bank. Much of this information is made conveniently accessible to scientists and engineers in printed form. The *CRC Handbook of Chemistry and Physics* includes a listing of the NSRDS publications in its appendices. Since 1972 much of the System's output has been presented in a quarterly called *Journal of Physical and Chemical Reference Data*. This is now a joint undertaking of NBS, American Chemical Society and the American Institute of Physics which increases the dissemination of the data. The ACS also supplies reprints with a charge. The journal provides critically evaluated physical and chemical property data with information on sources and criteria used for evaluation.

Another major data analysis center associated with NSRDS is the Center for Information and Numerical Data Analysis and Synthesis (CINDAS) located at Purdue University. One of the centers within CINDAS is the Thermophysical Properties Research Center. Its fourteen volume *Thermophysical Properties of Matter* is noted on the following list. Compilations of data from original sources like the NSRDS are known as secondary and tertiary sources. These include handbooks, journals and books which contain data compilations, reports and updates on projects and activities. Abstracting and indexing services, especially *Chemical Abstracts*, are valuable for locating various types of data.

Separate compilations of tables can be located in library card catalogs under the subject of the tables, i.e. Thermodynamics–Tables, etc.

A SELECTIVE LIST OF TABLES OR HANDBOOKS, ETC. CONTAINING TABLES

The choice of titles listed in this chapter and that on handbooks was arbitrary. Many of the titles in the following list could be classified as handbooks and many of the titles in the chapter on handbooks contain numerous tables.

Handbooks and Tables in Science and Technology (2nd ed.). Russell H. Powell (ed.), Oryx Press, Phoenix, 1983, 297 pp.

> Bibliography of handbooks and tables. Author and subject indexes enable the user to select the appropriate compilation or handbook. Brief descriptive notes accompany titles.

CRC Handbook of Tables for Applied Engineering Science. Ray E. Bolz and George L. Tuve (eds.), CRC Press, Boca Raton, Fla., 1973, 1166 pp.

> Provides access to minimum basic data in the various fields of engineering and references to more complete sources. Data is presented in metric as well as conventional units. Special attention given to environmental hazards and human safety. Index of properties including physical, chemical, electrical, mechanical.

Handbook of Chemistry and Physics: A Ready Reference Book of Chemical and Physical Data. CRC Press, Boca Raton, Fla. Annual. Published since 1913.

Data, including tables, supporting physics, chemistry and related sciences. Information on symbols and technology. Revised annually, thus offering most current data.

CRC Handbook of Mathematical Sciences (5th ed.). William H. Beyer (ed.), CRC Press, Boca Raton, Fla., 1978, 982 pp.

The 5th edition of this handbook has been "restructured in content and format" and contains much new material on such subjects as theory of numbers, differential equations, complex variables, special functions and astrodynamics.

CRC Standard Mathematical Tables (26th ed.). William H. Beyer (ed.), CRC Press, Boca Raton, Fla., 1981, 624 pp.

Includes integral tables, determinants and matrices and mensuration.

Composite Index for CRC Handbooks (2nd ed.). CRC Press, Boca Raton, Fla., 1977, 1111 pp.

Part 1: Subject entries. Part 2: Chemical substances entries. Enables users to locate tables in the CRC handbooks. Key to approximately 50 handbooks including the *Handbook of Chemistry and Physics*, on subjects such as environmental control, toxicology, lasers, materials, optics. Directs users to entries giving the title of the handbook and the page on which appropriate information appear.

Gas Tables: Thermodynamic Properties of Air, Products of Combustion and Component Gas: Compressible Flow Functions including those of Ascher H. Shapiro and Gilbert M. Edelman (2nd ed.). Joseph H. Keenan et al., John Wiley & Sons, New York, 1980, 217 pp.

Thermodynamic properties of air at low pressures and of each of its component gases have been reevaluated using the most current physical, molecular and spectroscopic constants. Thermodynamic properties for combustion products of hydrocarbon fuel with 100% theoretical air, useful to the energy industries, are included in the tables.

Handbook of Physical Calculations: Definitions, Formulas, Technical Applications, Physical Tables, Conversion Tables, Graphs, Dictionary of Physical Terms. Jan J. Tuma, McGraw-Hill Book Co., New York, 1976, 370 pp.

A functional handbook that contains definitions and formulas as well as tables relative to technological physics, emphasizing practical applications.

Handbook of Thermodynamics and Charts. Kuzman Ravnjevic, Hemisphere Publishing, Washington, D.C., 1976, 382 pp.

Comprehensive collection of thermodynamic or transport data for various substances in the solid, liquid or gaseous states.

Practicing Scientist's Handbook: A Guide for Physical and Terrestrial Scientists and Engineers. Alfred J. Moses, Van Nostrand Reinhold, New York, 1978, 1292 pp.

Comprehensive source of materials property data: physical, mechanical, chemical, thermodynamic, thermal, electrical and electronic, optical, acoustical and nuclear. Materials are divided into chemical elements, organic compounds, alloys, refractories and supercooled liquids, composites, polymers and adhesives, semi- and superconductors, and the environment.

Statistical Tables for Science, Engineering, Management and Business Studies (2nd ed.). John Murdoch and J. A. Barnes, Halsted Press, New York, 1977, 46 pp.

Commonly used statistical and mathematical tables including basic distribution and significance tables, statistical quality control tables, and accounting tables.

ASME Steam Tables (3rd ed.). C.A. Meyer et al., American Society of Mechanical Engineers, New York, 1977, 329 pp.

Thermodynamic or transport properties of steam shown in nine tables and thirteen charts computed from the equations adopted in the 1967 *IFC Formulation for Industrial Use.*

Thermophysical Properties of Matter: The TPRC Data Series: A Comprehensive Compilation of Data by the Thermophysical Properties Research Center (TPRC), Purdue University (Vols. 1–14). Y.S. Touloukian and C.Y. Ho, (eds.), Plenum Publishing Corp., New York, 1970–79.

Comprehensive collection of evaluated numerical data and experimental measurements on specific thermophysical properties. Volumes cover thermal conductivity, specific heat, thermal radiative properties, thermal diffusivity, viscosity, thermal expansion; metallic elements and alloys; non-metallic solids, liquids and gases. Vol. 14, *Master Index to Materials and Properties*, combines all the indexes of the other volumes.

International Critical Tables of Numerical Data, Physics, Chemistry and Technology (Vols. 1–7, index). National Research Council, McGraw-Hill Book Co., New York, 1926–33.

Will not be updated. Collections of data in above subject areas continue in issues of *Journal of Physical and Chemical Reference Data.*

Unit Conversions and Formulas Manual. Nicholas P. and Paul N. Cheremisinoff, Ann Arbor Science Publishers, Ann Arbor, Mich., 1980, 171 pp.

Handy reference of unit conversions and formulas for numerous engineering subjects including hydraulics, heat transfer, statics and dynamics.

Handbook of Mathematical Tables and Formulas (5th ed.). Richard S. Burington, McGraw-Hill Book Co., New York, 1973, 500 pp.

First part includes main formulas and theorems of algebra, geometry, trigonometry, calculus, vector analysis, sets, logic, matrices, linear algebra, numerical analysis, differential equations, some special functions, Fourier and Laplace transforms, complex variables and statistics; second includes tables of logarithms, trigonometry, exponential and hyperbolic functions, powers and roots, probability distributions, annuity, etc.

8

Encyclopedias

Most people encounter a reference tool called an encyclopedia quite early in their education, but relatively few know that there are many types of encyclopedias besides the general works frequently found in schools, homes, or public libraries. They can range from a single volume covering all fields of knowledge to a multi-volume set covering a very specialized subject like chemical technology.

The features that identify an encyclopedia are:

- Preparation by a team of experts;
- Articles signed by authors, frequently in alphabetical order, and ranging from a column to several pages in length;

- Frequent inclusion of bibliographies or reading lists.

Some encyclopedias have irregularly issued supplements, and a few have annual yearbooks.

It should be noted that some publications that meet the requirements listed above don't even have the word "encyclopedia" in their titles. A number of them may be called "handbooks" or "dictionaries." The chapters discussing these three formats have tried to clearly define the differences, and cite examples that support the definitions.

The publications described below are examples of: a single and a multivolume general encyclopedia; a single and a multivolume science/technology encyclopedia; and a few single and multivolume subject encyclopedias.

GENERAL

The New Columbia Encyclopedia (4th Ed.). William H. Harris and Judith S. Levey (eds.), Columbia Univ. Press, New York, 1975, 3052 pp.

> The best-known, most authoritative one-volume general encyclopedia. The 4th edition offers far more detailed articles on scientific processes and theories than did earlier editions (1935, 1950, and 1963). Very useful for factual and background information.

The Encyclopedia Americana. Americana Corporation, Danbury, Conn. Published since 1829, and annually since 1936.

> One of the best known, comprehensive general multivolume encyclopedias. The 1981 edition contains 27,000 pages, 52,800 articles, and 22,900 illustrations and maps in 30 volumes.
>
> Other comparable encyclopedias include *Encyclopaedia Britannica, Collier's Encyclopedia* and the new *Academic*

American Encyclopedia, first published in 1980. All of these employ the policy of continuous revision, involving annual updating in areas such as elections and political events. In the *Americana*, 6740 new or revised articles were prepared from 1978 through 1981. Many of these general multivolume sets have annual yearbook volumes that include the more recent articles until a revised edition is published.

SCIENCE AND TECHNOLOGY

Van Nostrand's Scientific Encyclopedia (6th Ed.). Van Nostrand Reinhold Co., New York, 1982, 3100 pp.

Contains over 7,300 alphabetically arranged entries in bioscience, chemistry, earth and atmospheric sciences, energy and power technology, mathematics, information sciences, medical sciences, physics, space and planetary sciences. Includes over 2500 illustrations and 600 tables. Available in one volume or a two-volume set.

McGraw-Hill Encyclopedia of Science & Technology (5th Ed.) (15 Vols.). McGraw-Hill Book Co., New York, 1982.

The major multivolume encyclopedia covering every field of science and technology. Includes 7,700 signed articles, 2,000 of which were revised or rewritten since the 4th edition (1977). Most articles include bibliographies. Volume 15 is the index, containing 150,000 entries for terms, concepts, and people. As noted below, some of the material from earlier editions of this work have been incorporated into subject-focused encyclopedias, published separately by McGraw-Hill.

Since 1962, the publisher has been producing the *McGraw-Hill Yearbook of Science & Technology* to assist in keeping up to date between editions.

SUBJECT

Encyclopaedic Dictionary of Physics (9 Vols.). J. Thewlis (ed.), Pergamon Press, Oxford, 1961–1964.

A scholarly multivolume encyclopedia on general, nuclear, solid-state, molecular, chemical, metal and vacuum physics; astronomy; geophysics; biophysics; and related subjects. The basic set is in seven volumes. Vol. 8 is a detailed index to the basic volumes, and Vol. 9 is a multilingual glossary in English, French, German, Spanish, Russian, and Japanese.

Supplementary volumes are being published to cover new and revised articles. The most recent is Vol. 5 (1975) but Vol. 6 has been announced for publication. Each has an index of terms, but as yet, there is no comprehensive index to the entire set.

Encyclopedia of Building and Construction Terms. Hugh Brooks, Prentice-Hall, Englewood Cliffs, N.J., 1983, 416 pp.

Almost 3,000 entries illustrate and describe tools, materials, pieces of equipment and new construction methods. The latest energy definitions and a comprehensive math manual of conversion factors, formulas, charts and tables are also included.

Encyclopedia of Chemical Technology (3rd Ed.). (Approx. 25 Vols.), Wiley-Interscience, New York, 1978–1984.

Familiarly known to many as Kirk-Othmer, after the editors of the earlier editions, this set is projected for completion in 1984 in 24 volumes and an index. About half of the articles deal with single chemical substances or groups of substances. There are also articles on industrial processes; uses; unit operations and processes in chemical engineering; and fundamentals. Temporary indexes published for Volumes 1–20, will be combined into one cumulated index volume when the set is completed.

Another important multivolume set in this general area is the *Encyclopedia of Chemical Processing and Design* (Marcel Dekker, New York). Begun in 1976, there are 16 volumes published so far.

Encyclopedia of Computer Science and Engineering (2nd Ed.). Anthony Ralston and Edwin D. Reilly (eds.), Van Nostrand Reinhold Co., New York, 1983, 1678 pp.

Contains 550 articles on hardware, systems, information and data, software, computing, and applications. Each article includes cross references. Also contains a number of useful appendices: abbreviations, computer science journals, departments of computer science, and a glossary of major terms in five languages. Includes a 5000 term index.

Encyclopedia of Computer Science and Technology (Approx. 20 Vols.). Jack Belzer (ed.), Marcel Dekker, New York, 1975–1984.

A major comprehensive encyclopedic survey of computer science addressed to the broad spectrum of its component subfields. Fifteen volumes have been published so far, with at least five more to come.

Encyclopedia of Wood: Wood as an Engineering Material. Sterling Publishing Co., New York, 1980, 376 pp.

A paperback reprint of the U.S. Department of Agriculture's *Wood Handbook*, 1974 revised edition. Note also the *Wood Engineering Handbook*, cited in the chapter on handbooks.

Illustrated Encyclopedia of Space Technology: A Comprehensive History of Space Exploration (1st U.S. Ed.). Kenneth Gatland, Harmony Books, New York, 1981, 289 pp.

Includes articles by fifteen authors who are engineers or science writers with connections to the American Institute of Aeronautics and Astronautics or the British Interplanetary Society. Excellent color illustrations.

International Petroleum Encyclopedia. Pennwell Publishing Co., Tulsa, Oklahoma, 1983, 450 pp.

A thorough encyclopedia covering the subject, including large sections on the role of state-owned oil companies, gas processing, synfuels, current exploratory regions. Its major feature is a 162-page worldwide petroleum atlas section, with maps showing oil and gas fields, refineries, and pipelines. Does not have a term index. Prepared by the publishers of *Oil & Gas Journal.*

McGraw-Hill Encyclopedia of Environmental Science (2nd Ed.). McGraw-Hill Book Co., New York, 1980, 858 pp.

Contains more than 250 articles, organized in two sections: A section containing five feature articles on topics of broad, general interest (e.g. Urban Planning, Environmental Analysis), and a section of alphabetically arranged articles dealing with environmental science and engineering. Some articles were taken from the 4th edition (1977) of the *McGraw-Hill Encyclopedia of Science and Technology*. Includes a detailed analytical index.

A companion volume, very similar in format, is the *McGraw-Hill Encyclopedia of Energy* (2nd edition, 1981).

9

Dictionaries

Scientific and engineering dictionaries are basically of two forms: those that provide meanings of terms, and those that give equivalents of terms in other languages.

The entries given in definition dictionaries are usually short, frequently two or three lines, and sometimes include illustrations or diagrams. These dictionaries can range from works covering many disciplines to those restricted to one specific subject, and are often single volumes.

The distinction between dictionaries, encyclopedias, and handbooks is frequently blurred by the publication of encyclopedic dictionaries and term books that deviate from the strict alphabetical arrangement of most dictionaries. The *Construction Glossary* and the *Encyclopaedic Dictionary of Mathematics for*

Engineers and Applied Scientists, cited below, are examples that might easily have been included in other chapters.

Foreign-language dictionaries, which may be bilingual or multilingual, rarely provide definitions. A bilingual dictionary is one that gives equivalents in two languages—usually English and one foreign language. Some go from the foreign language to English, others from English to the foreign language, and many will do both.

Multilingual, or polyglot, dictionaries usually cover English and up to six or seven foreign languages.

As noted in other chapters, the list of dictionaries given below is highly selective, with emphasis on recently published volumes.

DEFINITION DICTIONARIES

Computer Acronyms, Abbreviations, Etc. Claude P. Wrathall, Petrocelli Books, New York, 1981, 483 pp.

> Lists the meanings of more than 10,000 acronyms, abbreviations and names from the computer and communication field. Entries give only the explanation of the initialism, so most are only two or three words.

Construction Glossary: An Encyclopedic Reference and Manual. J. Stewart Stein, Wiley Interscience, New York, 1980, 1013 pp.

> The volume is organized into divisions and sections based on the MASTERFORMAT of the Construction Specifications Institute. Examples of the divisions used: site work; wood and plastics; doors and windows; conveying systems. Several useful appendixes, including weights and measures, and abbreviations for construction terms.

Dictionary of Civil Engineering (3rd ed.). J.S. Scott, Halsted Press, New York, 1981, 308 pp.

Covers terms in the broad field of civil engineering, except for building terms, which are included in a companion volume, *Dictionary of Building* (1964).

Dictionary of Electrical Engineering (2nd ed.). K.G. Jackson and R. Feinberg, Butterworth Publishers, Woburn, Mass., 1981, 356 pp.

Clear, concise definitions of terms currently in use in the United States and Great Britain. Includes some line drawings and cross-references.

Butterworth publishes a series of engineering term dictionaries. Others are:

Dictionary of Data Processing (2nd ed.) (1981).

Dictionary of Mechanical Engineering (2nd ed.) (1975).

Dictionary of Waste and Water Treatment (1981).

Dictionary of Audio, Radio and Video (1981).

Dictionary of Energy Technology (1982).

Dictionary of Telecommunications (1981).

Dictionary of Electronics (1981).

Dictionary of Geotechnics (1982).

Dictionary of Energy. Malcolm Slessor (ed.), Schocken Books, New York, 1982, 299 pp.

Standard dictionary arrangement. Each term is defined briefly, often followed by a longer, more detailed explanation. Includes diagrams, tables and cross-references.

A Dictionary of Named Effects and Laws in Chemistry, Physics and Mathematics (4th ed.). D.W.G. Ballentyne and D.R. Lovett, Chapman and Hall, New York, 1980, 346 pp.

Effects, laws, theories, and relations that are best known by the name of the person discovering or propounding them. Examples: Born-Haber Cycle; Euler's Transformation; Thomson Scattering.

Electronics Dictionary (4th ed.). John Markus, McGraw-Hill Book Co., New York, 1978, 745 pp.

"Accurate, easy-to-understand and up-to-date definitions for 17,090 terms used in solid-state electronics, computers, television, radio, medical electronics, industrial electronics, satellite communications and military electronics."

Includes an Electronics Style Manual, which summarizes troublesome spelling and grammatical problems encountered in writing about electronics.

Encyclopaedic Dictionary of Mathematics for Engineers and Applied Scientists. I.N. Sneddon (ed.), Pergamon Press, Elmsford, N.Y., 1976, 800 pp.

Typical dictionary arrangement, but some articles are fairly long (2000 to 3000 words), include bibliographies, and are signed. Includes a detailed subject index to topics not given separate entries.

Energy Deskbook. Samuel Glasstone, U.S. Dept. of Energy, Oak Ridge, Tenn., 1982, 453 pp. Available from NTIS as DE82013966.

Designed to serve as a convenient reference to energy-related terms and descriptions of current and potential energy sources. Entries vary in length from a few lines to several pages. Table of contents is alphabetical, and serves as an index.

Jane's Aerospace Dictionary. Bill Gunston, Jane's Publishing Company, New York, 1980, 483 pp.

A comprehensive guide to the language of aviation and space technology. Contains over 15,000 terms from astronautics, civil and military aeronautics, meteorology, materials science, and electronics.

The Masonry Glossary. International Masonry Institute, CBI Publishing Company, Boston, 1981, 144 pp.

Brief definitions of terms currently used in masonry construction. Includes frequent line drawings.

McGraw-Hill Dictionary of Scientific and Technical Terms (2nd ed.). Daniel N. Lapedes (ed.), McGraw-Hill Book Co., New York, 1978, 1750 pp.

Clear, concise definitions for over 100,000 terms in all areas of science and technology. Each definition includes an indication of the field that the term is used in. Contains about 3000 illustrations.

Robotics Sourcebook and Dictionary. David F. Tver and Roger W. Bolz, Industrial Press, New York, 1983, 258 pp.

A reference book covering most of the key aspects of current industrial robotics. Divided into four sections:

1. Introduction and dictionary of types.
2. Robotics dictionary of applications.
3. Robotics glossary and computer-control terminology.
4. Robotics manufacturers and typical specifications.

FOREIGN-LANGUAGE DICTIONARIES

Dictionary of Engineering and Technology (4th ed.). (2 volumes). Richard Ernst, Oxford University Press, New York, 1975–1980.

A multivolume work considered an authoritative source for translation of technical English and German. *Volume 1: German–English* contains 150,000 entries for German words and phrases. Volume 2 covers English–German.

Dictionary of Water and Sewage Engineering (2nd rev. ed.). Fritz Meinck and Helmut Möhle, Elsevier Scientific Publishing Co., New York, 1977, 737 pp.

A multilingual dictionary originally published in Germany. Part 1 is an alphabetical list of German terms, each numbered and accompanied by English, French and Italian equivalents. Part 2 comprises English, French and Italian terms in separate lists, cross-referenced by number to the German term in Part 1.

Elsevier's Dictionary of Metallurgy and Metal Working in Six Languages: English/American, French, Spanish, Italian, Dutch and German (2nd ed.). W.E. Clason (comp.), Elsevier Scientific Publishing, New York, 1978, 848 pp.

One of Elsevier's series of multilingual scientific and technical dictionaries. Others include:

Elsevier's Dictionary of Aeronautics (1964).

Elsevier's Dictionary of Automobile Engineering (1977).

Elsevier's Dictionary of Building Construction (1959).

Elsevier's Dictionary of Building Tools and Materials (1982).

Elsevier's Dictionary of Chemical Engineering (1969) (2 vols.).

Elsevier's Dictionary of Electronics and Waveguides (1966).

Elsevier's Dictionary of Measurement and Control (1977).

Elsevier's Dictionary of Nuclear Science and Technology (1970).

Elsevier's Electrotechnical Dictionary (1965).

Elsevier's Telecommunication Dictionary (1976).

The Multilingual Computer Dictionary. Alan Isaacs (ed.), Facts on File, New York, 1981, 332 pp.

Gives the English, German, Spanish, Italian, and Portuguese equivalents of 1600 computer terms.

Russian-English Scientific and Technical Dictionary (1st ed.) (2 vols.). M.H.T. and V.L. Alford, Pergamon Press, Elmsford, N.Y., 1970, 1423 pp.

Contains more than 100,000 entries in all fields of science and technology. Entries go from Russian–English only.

Technical Dictionary: Electricity, Mechanics, Mining, Metallurgy, Sciences: English-French, French-English (7th ed.). Francis Cusset, Chemical Publishing Co., New York, 1967, 434 pp.

A simple glossary with no introductory information. Includes a brief appendix for conversion data to and from metric measures.

10

Directories (People, Organizations and Products)

Directories, as defined here, are reference tools that contain information about people, companies, associations, products, services, etc. They are usually separate bound books, but can appear in other formats as well. For example, some are published as special issues of periodicals, and may contain an organization's membership directory, or a buyer's guide to manufacturers and products. Examples of the latter are included in Chapter 15 on "Trade Literature."

Directories that are produced primarily for one function may also be useful in another capacity. An obvious example of this is the telephone book, which provides certain information on individuals and companies, while its yellow pages are a directory of products and services.

"People" directories are more frequently called biographical dictionaries, or "Who's Who's." Within their established limitations, they provide short biographical sketches, usually including age, family history, education, occupational history, memberships, awards, important publications, and current address.

These publications vary greatly in scope and reliability. Some cover subject areas (*Who's Who in Engineering*), others cover geographic regions (*Who's Who in the East*). The directories described in this chapter are published by respected, well-known publishers, and have clearly defined criteria for inclusion of biographies. The reader should be aware, however, that there are some biographical directories on the market which could be called "vanity" publications. In these volumes, there doesn't seem to be any clear standard for inclusion (other, perhaps, than a willingness by the prospective biographee to buy a copy of the volume).

Organization directories are numerous and variable. Some cover specific subject areas (*World Aviation Directory*), others specific types of organizations (*Research Centers Directory*). There are, in fact, so many directories now available that Gale Research Company has just published in 1982 the 2nd edition of a *Directory of Directories*, which has 1000 pages and over 7000 entries.

There are also directories whose primary emphasis is on products or services. These are particularly valuable for an engineer who finds that he needs to know something about all the companies that make a particular piece of equipment, or information on a number of firms that have expert consultants in an unfamiliar area.

Directories have traditionally been printed and published, but recently a number of them have become available online in a computer-searchable form. These offer certain advantages over the printed versions:

1. Most are updated frequently so they don't get out of date;

2. For the occasional user, it could be considerably less expensive to search a directory online than to buy a print copy.

What follows here is a highly selective list of directories that have broad use, and should be of value to all kinds of engineers. As noted above, a comprehensive list is the subject of a whole separate book. Those that are available both in printed and online form, as well as those that are only in one form or the other, are included where appropriate.

BIOGRAPHICAL DIRECTORIES

American Men & Women of Science. Jaques Cattell Press (ed.), R.R. Bowker, New York. Published irregularly since 1906.

> Now published every three years. The 15th edition, in seven volumes, is dated 1982. This edition profiles 130,000 living scientists in the physical and biological fields, as well as public health scientists, engineers, mathematicians, statisticians and computer scientists.
>
> Available for computer-searching from BRS and Dialog Information Services.

Marquis Who's Who Publications: Index to All Books. Marquis Who's Who, Inc., Chicago. Published annually since 1974.

> Marquis is the foremost publisher of biographic directories in the U.S. Among their publications are: *Who's Who in America, Who Was Who in American History–Science & Technology*, and *World Who's Who in Science*. This particular book is a cross-reference to over 300,000 names appearing in the current editions of fourteen of their publications.
>
> The *Who's Who in America* publication is also available for computer-searching from Dialog Information Services.

Who's Who in Engineering. American Association of Engineering Societies, New York. Published since 1970.

The 5th edition is dated 1982. The first and second editions were titled *Engineers of Distinction*. Most of the volume contains biographies, but the first 48 pages give information on engineering societies and awards. Also has specialization and geographical indexes.

Who's Who in Science in Europe (3rd ed., 4 vols.). Francis Hodgson Books, Ltd., Guernsey, U.K., 1978, 3534 pp.

Includes 50,000 entries in natural and physical sciences, medicine, and agriculture. Previous editions were dated 1967 and 1972.

Who's Who in Technology Today (3rd ed., 4 vols.). J. Dick & Co., Highland Park, Ill., 1982, 3320 pp.

This edition contains biographies of 27,000 engineers, scientists, inventors and managers. There are four volumes, as follows:

Vol. 1: Electronic and physics technologies.

Vol. 2: Mechanical and earth science technologies.

Vol. 3: Chemical and bioscience technologies.

Vol. 4: Index (by name and principal expertise).

ORGANIZATION DIRECTORIES

Encyclopedia of Associations. Gale Research Co., Detroit. Published since 1956, and annually since 1975.

A comprehensive guide to national and international organizations, societies, and associations. For a more detailed description, see Chapter 18 on "Professional Societies".

Available for computer-searching from Dialog Information Services.

Industrial Research Laboratories of the United States. Jaques Cattell Press (ed.), R.R. Bowker Co., New York. Published since 1920, and every two or three years since 1975.

A guide to research and development activities conducted by nongovernmental or nonacademically supported organizations. Over 11,000 listings are included in the 18th edition (1983). Includes a geographic index, a personnel index, and a subject index to activities.

Research Centers Directory. Gale Research Co., Detroit. Published since 1960.

"A guide to university-related and other nonprofit research organizations established on a permanent basis and carrying on continuing research programs in agriculture, business, conservation, education, engineering and technology, government, law, life sciences, mathematics, area studies, physical and earth sciences, social sciences, and humanities."

The 8th edition (1983) contains 6314 entries. Supplements, called *New Research Centers*, are published between editions.

Gale now is publishing a companion volume, *Government Research Centers Directory*. The 2nd edition is dated 1982.

The United States Government Manual. U.S. Government Printing Office, Washington, D.C. Published annually since 1935.

Describes the purposes and programs of most agencies and lists top personnel. Detailed information on

Congressmen and their staffs may be found in the *Congressional Directory*, which is published for each session of Congress by the Government Printing Office.

PRODUCTS AND SERVICES DIRECTORIES: PRINTED

Consultants and Consulting Organizations Directory. Gale Research Co., Detroit. Published irregularly since 1966.

"A reference guide to concerns and individuals engaged in consultation for business and industry." The 5th edition (1982) provides details about 7064 firms, individuals, and organizations.

Gale also publishes two companion publications: *New Consultants*, which supplements the above directory between editions; and *Who's Who in Consulting*, which gives biographical data on over 3000 individuals.

Energy-Efficient Products and Systems: A Comparative Catalog for Architects and Engineers (1st ed.). John Wiley & Sons, Inc., New York, 1983, looseleaf.

This directory contains about 400 product entries, representing over 260 manufacturers. It covers a wide range of building products, building envelope systems, mechanical systems, and mechanical system components. It is divided into four sections:

The Building Envelope;

Mechanical Systems;

Domestic and Process Hot Water;

Electrical Systems.

A subscription updating service is available.

MacRAE's Blue Book. MacRAE's Blue Book, Inc., Plainview, N.Y. Published annually since 1893.

One of the oldest industrial directories, with over 18,000 product headings and more than 50,000 manufacturers listed. Presently in 5 volumes, as follows:

Vol. 1: Corporate Index.

Vols. 2–4: Products and Services Catalogs.

Vol. 5: Manufacturer's Catalog.

This publisher also publishes separate industrial directories for each state. These are revised annually. Each one has three basic arrangements

- An alphabetical list by firm name
- A geographical listing by counties and regions
- A classified index by product

Sweet's Catalog File: Products for Industrial Construction and Renovation. McGraw-Hill, New York. Published annually since 1906.

A multivolume annual cumulation of manufacturers' catalogs. The 1981 edition of this particular file is in four volumes. Its main divisions are general data, sitework, concrete, masonry, metals, wood and plastics, thermal and moisture protection, doors and windows, finishes, specialties, equipment, furnishings, special construction, conveying systems, mechanical, electrical.

Sweet's Catalog File: Products for Engineering–Mechanical, Electrical, Civil and Related Products. McGraw-Hill, New York. Published annually since 1906.

Another in the six Sweet's Catalog File Series. The 1981 edition of this one is in ten separate volumes. The same divisions are used for all.

Thomas Register of American Manufacturers and Thomas Register Catalog File. Thomas Publishing Company, New York. Published annually in multiple volumes since 1905.

The 73rd edition (1983) consists of 17 volumes:

Vols. 1–9: Products and Services.

Vols. 10–11: Company Profiles (Alphabetical).

Vols. 12–17: Catalogs of Companies.

PRODUCTS AND SERVICES DIRECTORIES: ONLINE

Electronic Yellow Pages

A series of separate computer-search data bases produced by Market Data Retrieval, Inc., available through Dialog Information Services. Data are taken from the yellow pages of over 4800 telephone books and special directories. Each is described briefly below.

Construction Directory

Contains mailing and descriptive information on contractors and construction agencies, including housing contractors; industrial builders; highway, bridge, and tunnel construction companies; companies that specialize in concrete work, structural steel, excavating, and foundations; and wrecking and demolition companies. Also covers such related trades as plumbing; heating; painting; paper handing; masonry; plastering; terrazzo, tile, and marble work; flooring; and roofing.

Financial Services Directory

Contains mailing and descriptive information on more than 90,000 banks, savings and loan companies, and credit unions.

Manufacturers Directory

Contains mailing and descriptive information on more than 540,000 U.S. factories, plants, mills, and other companies that transform materials and substances into new products. Included are producers of such products as petroleum, chemicals, and plastics; lumber, furniture, and paper; apparel and textiles, leather, stone clay, glass, and concrete; fabricated metal products and machinery; electrical, electronic, and transportation equipment; and food, beverage, and tobacco products. These companies belong to the Standard Industrial Classification (SIC) codes 1850, 2000–3999.

Professionals Directory

Contains mailing and descriptive information on over one million professionals in the fields of insurance, real estate, medicine, law, engineering, and accounting. Also included are hospitals, medical laboratories, and clinics.

Retailers Directory

Consists of three files of mailing and descriptive information on more than three million retail businesses, covering Standard Industrial Classification (SIC) codes 5200–5499, 5800–5999. Includes lumber stores; paint stores; hardware stores; retail nurseries and garden supply stores; grocery, meat, and other food stores; gasoline stations; automobile and other vehicle dealers; clothing stores; furniture stores; book stores; eating and drinking establishments; and florists.

Services Directory

Contains mailing and descriptive information on over 1.9 million businesses in the service industries. Covers businesses in the

Standard Industrial Classification (SIC) codes 7000–7999. Includes hotels, motels, and resorts; cleaners and dyers; direct mail and advertising services; commercial photography, art, graphics, stenographic, and reproduction services; employment agencies and temporary help supply services; computer programming, software, and data processing services; management, consulting, and public relations services; detective agencies and protective services; equipment rental and leasing services; photofinishing laboratories; repair shops; sports clubs; motion picture theaters; and amusement and recreation centers.

Wholesalers Directory

Contains mailing and descriptive information on more than 900,000 wholesale dealers. Covers dealers of such products as automotive parts; office furniture and equipment; housewares; lumber and plywood; toys and hobby supplies; electrical equipment; plumbing and heating equipment; and farm and garden machinery.

TECHNET

A data base providing information on specifications, standards, and products. It is produced by and is available only from Information Handling Services.

The *Vendor Product Data* file contains references to more than 10 million products listed in over 24,000 industrial vendor catalogs. Includes company name, address, telephone number and worldwide sales office locations.

A complete description of this data base and other IHS services is included in the chapter on *Standards and Specifications.*

11

Standards and Specifications

It is inevitable that at some time in their careers, engineers will have to deal with specifications, standards, codes, and similar documents. This chapter attempts to bring some order into one of the most complex areas of engineering literature.

The words "standard" and "specification" are sometimes used interchangeably, but there are differences in the definitions of the terms.

Standards in general refer to rules, techniques and conditions which must be adhered to in engineering design, industrial practices, units of measurement, terminology, and so on. They may be voluntary or mandatory, depending on the country or the particular field involved.

Engineering standards may be thought of as rules for the uniformity, size, quality, performance, shape, definition and testing methodology of manufactured products. These are usually voluntary and are called *consensus standards*, because diverse groups, such as manufacturers, merchandisers, experts, and consumers, must agree on the requirements and the intended result.

A *specification*, according to the American National Standards Institute (ANSI) is "a concise statement of the requirement for a material, process, method, procedure or service, including, whenever possible, the exact procedure by which it can be determined that the conditions are met within the tolerances specified in the statement; a specification does not have to cover specifically recurring subjects or subject of wide use, or even existing objects." To an engineer, *specification* refers to a document that contains the requirements for materials, products, or services that are to be purchased by an industry or government agency.

The applicability of a specification is usually more limited than that of a standard. ANSI describes the relationship as follows: "A standard is a specification accepted by recognized authority as the most practical and appropriate current solution of a recurring problem."

TYPES OF ENGINEERING STANDARDS

Standards and specifications fall into the following categories, based on their purpose:

1. *Dimensional*: To assure uniformity and interchangeability of the types and sizes of manufactured components.
2. *Qualitative*: To specify minimum performance, quality, composition and properties of materials.
3. *Test Methods*: To provide uniform procedures for comparing the quality of products.

4. *Codes of Practice*: To define correct procedures for construction, operation, installation, and maintenance for uniformity and safety.
5. *Standard Definitions and Symbols*: To secure precision and consistency in the use of words, phrases, and symbols.
6. *Documentation*: To provide standard rules for authors, and procedures for format of technical publications.

FORMULATORS OF STANDARDS

Standards and specifications may originate with a number of sources.

Industry

Most companies have some internal standards of the types described above. These documents are usually not intended for other than company use, and may even be restricted or confidential.

Professional Associations

Many technical societies produce standards, codes, and specifications for the industrial communities they serve. Notable among them are the American Society for Testing and Materials (ASTM), the Society of Automotive Engineers (SAE), and the Aerospace Industries Association (AIA), which account for over half of the consensus standards produced in the United States.

ASTM deserves special attention as the largest society solely dedicated to standardization. It was established in 1898 for "the development of standards on characteristics and performance of materials, products, systems, and services; and the promotion of related knowledge."

ASTM standards—test methods, specifications, definitions, practices, and classification—are written by those having expertise in specific areas, who choose voluntarily to work within the ASTM system. Current membership is over 28,000 organizations and individuals, worldwide, with a total unit participation of well over 80,000 in 138 technical committees. In addition to the *Annual Book of ASTM Standards*, ASTM also provides numerous other technical publications and related material which have evolved from committee activities. Major ASTM publications include:

- *Annual Book of ASTM Standards*. The 1983 edition contains more than 6700 standards. It is published in 66 volumes, totaling 55,000 pages.
- *Compilations of ASTM Standards*. These are special groupings of related standards from one or more volumes of the *Annual Book of ASTM Standards*. Some of the topics are listed below:

Building codes	Petroleum products
Concrete	Quality control
Electrical conductors	Solar energy
Fire testing	Spectrochemical analysis
Magnetic testing	Structural steel
Ferroalloys	Thermocouples

  Such volumes may be particularly useful to practicing engineers, and are much less expensive than the entire set of standards.
- *Special Technical Publications. (STP's)* STP's are books devoted to specific technical, scientific, and

standardization topics. Most are based on symposia sponsored by ASTM technical committees.

- *ASTM Journal of Testing and Evaluation*. bimonthly periodical, published since 1966.
- *Cement, Concrete, and Aggregates*. semiannual periodical, published since 1979.
- *Geotechnical Testing Journal*. quarterly periodical, published since 1978.
- *Composites Technology Review*. quarterly periodical, published since 1979.
- *Journal of Forensic Sciences*. quarterly periodical, published since 1956.
- *ASTM Standardization News*. monthly periodical, published since 1973.

Copies of standards and other publications may be obtained from:

American Society for Testing and Materials
1916 Race Street
Philadelphia, Pa. 19103
(215) 299-5413.

Government Agencies

In most countries, such agencies are among the largest originators of standards and specifications. Three agencies in the U.S. produce the great majority of governmental standards: The National Bureau of Standards (NBS), the Department of Defense (DOD), and the General Services Administration (GSA).

The NBS was established in 1901 to develop standards for temperature, electricity, mass and length. It now deals with all phases of research and development in physics, mathematics, chemistry and engineering to improve standards and methods of measurement.

The Bureau generates many publications, including Monographs, National Standard Reference Data Series, Building Science Series, and Voluntary Product Standards. Its major periodical is the *Journal of Research of the NBS*, published since 1901, sometimes in several parts. Since 1977, it is again being published as one title, six times a year.

One specialized activity within NBS could be especially useful for engineers working with standards: The National Center for Standards and Certification Information. The services of this agency will be discussed in some detail in the next section of this chapter, "Locating Information on Specifications and Standards."

Other governmental specifications and standards may be divided into two categories: military and nonmilitary. The Department of Defense has over 39,000 standards, specifications, military sheet form standards, Air Force-Navy aeronautical standards, qualified product lists, and handbooks currently in force. Over 5000 current Federal standards, specifications, commercial item descriptions, and qualified product lists are coordinated by the General Services Administration.

National Standards Organizations

Many countries have a national agency to coordinate standardization activities. Among those that engineers are most likely to encounter are: The American National Standards Institute (ANSI); the British Standards Institution (BSI); the Canadian Standards Association (CSA); and the Deutsches Institut für Normung e. V. (DIN). Information on these and the organizations of 55 other countries may be found in:

Use of Engineering Literature. K. W. Mildren (ed.), Butterworths, London, 1976, pp. 100–109.

ANSI is the coordinator of America's voluntary standards system. The system meets national standards needs by marshalling the competence and cooperation of commerce and industry, standards developing organizations, and public and consumer interests. ANSI has several major functions:

- Coordinating the voluntary development of national consensus standards;
- Approving standards as American National Standards;
- Managing and coordinating U.S. participation in the work of nongovernmental international standards organizations;
- Serving as a clearinghouse and information center for American National Standards and international standards.

ANSI publishes an annual catalog of all approved standards, and periodic supplements. In addition, the Institute publishes a biweekly periodical, *Standards Action*, which solicits comments on standards ANSI is considering for approval and reports on final actions and newly published standards. A biweekly newsletter–*ANSI Reporter*--reports on policy-level actions of ANSI and the international organizations to which it belongs and on standards-related actions and proposals of the U.S. government.

International Agencies

There are a number of such organizations. The two that engineers are likely to encounter are the International Organization for Standardization (ISO), and the International Electrotechnical Commission (IEC), both headquartered in Geneva, Switzerland.

The ISO was established in 1947 to succeed various international bodies dating back to 1926. It is a non-governmental organization with consultative status to the United Nations, and presently has 89 members. ISO publishes international standards and technical reports on all subjects except those concerned with the electrical and electrotechnical industries, which are handled by IEC. ISO standards are listed in their own catalog, and also in the ANSI catalog, cited below.

The IEC was founded in 1906 to promote cooperation in the electrotechnical industry. It presently has 43 national committees. Among IEC's accomplishments are the development of a multilanguage vocabulary with more than 100,000 terms, and origination of the International System (SI) of units of measurement. IEC standards are also included in the ANSI catalog.

LOCATING INFORMATION ON SPECIFICATIONS AND STANDARDS

Engineers today are fortunate to have several approaches to locating standards information, including the traditional printed indexes, several very new computer–searchable data bases, and a specialized service of the National Bureau of Standards.

Printed Indexes

Catalog of American National Standards. American National Standards Institute, New York. Published annually since 1923.

> As noted earlier in the chapter, this catalog lists all ANSI-approved standards. It also includes information on ISO and IEC standards and special publications. Bimonthly supplements.

Index and Directory of U.S. Industry Standards (2 Vols.). Information Handling Services, Englewood, Colorado, 1983, 2000 pp.

This is a first attempt to provide printed access to a majority of current U.S. industry standards. Covers over 20,000 documents from 38 organizations. It also provides addresses and phone numbers of over 400 standards-preparing bodies, and has a section that gives current ANSI numbers for documents that pre-date 1978 when the ANSI designation system was completely revised.

Index of Federal Specifications, Standards and Commercial Item Descriptions. U.S. General Services Administration, U.S. Government Printing Office, Washington, D.C. Published annually since 1952.

This work lists nonmilitary specifications accepted for federal use. It is arranged by subject classification, by number, and alphabetically by title, and includes price, date, edition number, and so forth. Bimonthly supplements.

Index of Specifications and Standards. U.S. Department of Defense, Washington, D.C. Published annually since 1951.

Issued with cumulative bimonthly supplements, this publication lists all unclassified specifications and standards adopted by the Department of Defense. One volume is arranged alphabetically by title, one by a classification system, and one by the standard numbers. Availability is indicated, as well as number and date of latest edition, and so on.

World Industrial Standards Speedy Finder. International Technical Information Institute, Tokyo, 1983. 750 pp. Distributed in U.S. and Canada by ASTM.

A cross-reference to over 53,000 standards from the United States, Great Britain, West Germany, France, Japan, and the International Standards Organization. Also includes safety standards from Canada and Australia.

A valuable feature is that when one country's standards on a subject are located, all related standards of the other countries can be found in the same place.

Computer-Search Data Bases

Data bases devoted to standards and specifications are a relatively recent development in computer-search services. There are presently four such data bases, produced by two companies that offer a wide variety of services in this specialized area of information: Information Handling Services and the National Standards Association. Further details about these companies and their services will be included in the section on "Obtaining Copies of Standards and Specifications."

INDUSTRY AND INTERNATIONAL STANDARDS

Producer: Information Handling Services (IHS)
Online Service: Bibliographic Retrieval Services (BRS)
Content: Contains citations to over 70,000 standards produced by U.S. and international agencies and industry standards societies. Also contains references to standards of all societies included in the National Bureau of Standards voluntary engineering standards database. References can be retrieved by document number, title, subject classification (IHS locator code), society, document type, and American National Standards Institute (ANSI) approval indicator. Each record contains a reference to the cartridge frame and location in IHS's VSMF (Visual Search Microfilm System), which contains the full text of the document.
Language: Primarily English, French, and German
Coverage: International
Time Span: All currently active standards
Updating: Every 60 days

MILITARY & FEDERAL SPECS & STANDARDS

Producer: Information Handling Services (IHS)
Online Service: BRS
Content: Contains citations to over 196,000 active and historical non-classified standards and specifications, includes references to Military Standards and Specifications, Federal Standards and Specifications, Joint Army-Navy Specifications, Military Standard Drawings, and Qualified Products Lists (QPLs). References can be retrieved by document number, title, Federal Supply Classification (FSC) code, and subject classification (IHS locator code). Each record contains a reference to the cartridge frame and location in IHS's VSMF (Visual Search Microfilm System), which contains the full text of the document.
Language: English
Coverage: U.S.
Time Span: 1964 to date
Updating: Semimonthly

STANDARDS & SPECIFICATIONS

Producer: National Standards Association, Inc.
Online Service: DIALOG Information Services
Content: Contains citations to government and industry standards, specifications and related documents. Covers over 39,000 U.S. Department of Defense standards, specifications, military sheet form standards, Air Force-Navy Aeronautical Standards (ANs), Qualified Products Lists (QPLs), and handbooks, including 1400 NATO and related standards; 6,500 U.S. General Services Administration (GSA) standards, specifications, Commercial Item Descriptions (CIDs) and QPLs; and 27,300 U.S. standards developed by 432 private organizations such as the American Society for Testing and Materials,

Society of Automotive Engineers, Aerospace Industries Association, American National Standards Institute, and Underwriters Laboratories. Each record includes a title and identification number identifying the standard or specification, its issuing organization, Federal Supply Classification code, and whether the document has been cancelled or superseded. A notation is included if designated as an American National Standard and, for international standards, if approved by the U.S. Also includes vendors of products and services conforming to the standard or specification.
Language: English
Coverage: U.S.
Time Span: Covers latest issue of all standards and specifications, most of which have been issued since 1950 (earliest data from 1920).

TECH-NET

Producer: Information Handling Services (IHS)
Online Service: Information Handling Services
Content: Provides access to commonly used industry standards and codes, military specifications and government documents, and industrial product information.

- *Industry Codes and Standards*. Contains references to commonly used standards from U.S. and international agencies and industry standards societies, e.g., American National Standards Institute, American Society for Testing and Materials, Society of Automotive Engineers, Underwriters Laboratories, Inc., British Standards Institute, International Organization for Standardization, Japanese Industrial Standards.
- *Government Specifications*. Contains references to military and federal specifications and standards and to military standard drawings.

- *Vendor Product Data.* Contains references to more than 10 million products listed in over 24,000 industrial vendor catalogs. Includes company name, address, telephone number and worldwide sales office locations.

Each product description, code, standard, and specification contains a reference to the cartridge and frame location in IHS's VSMF (Visual Search Microfilm System), which contains the full text and/or full catalog data.
Language: English
Coverage: International
Updating: Varies, from daily to semiannually

National Bureau of Standards, National Center for Standards and Certification Information (NCSCI)

This agency is an information center for standards-related information. It was established in 1965 under its former name, the Standards Information Service. Its objective is to respond to the needs of government, industry and the general public for information on domestic and foreign standards, regulations, certification, and standards-related activities. NCSCI does not supply copies of standards, but rather responds to inquiries on the existence, source, and availability of standards and related documents, and prepares lists and indexes.

The NCSCI reference collection includes over 240,000 standards, specifications, and like material from U.S. organizations, state purchasing offices, government agencies, and major foreign and international standardizing bodies. The Center responds to over 5,000 individual inquiries annually, and publishes semiannually a *KWIC (Key-Word-In-Context) Index of U.S. Voluntary Engineering Standards.* This index is available in forms varying from microfiche to computer printout, at prices currently ranging from $15–200.

The NCSCI may be reached at the following address and phone number:

National Center for Standards and Certification Information
Room B166, Technology Building
National Bureau of Standards
Washington, D.C. 20234
(301) 921-2587

OBTAINING COPIES OF STANDARDS AND SPECIFICATIONS

There are a variety of ways of acquiring copies of standards. They range from mail orders to telephone to telex to online orders, depending on how urgent the need is.

Single copies of military specifications and standards may be obtained without charge from:

Naval Publications and Forms Center
5801 Tabor Avenue
Philadelphia, Pa. 19120

All of these military documents, as well as most of the other standards included in the above-mentioned printed and online indexes, may be acquired through any of four commercial companies. Their addresses, phone numbers, and summaries of services are listed below.

Global Engineering Documents

Global Engineering Documents
2625 Hickory Street

P.O. Box 2504
Santa Ana, Calif. 92707
(800) 854-7179

Global will supply paper copies of a wide variety of standards, including those in the BRS data bases INDUSTRY AND INTERNATIONAL STANDARDS and MILITARY AND FEDERAL SPECIFICATIONS AND STANDARDS.

Information Handling Services (IHS)

Information Handling Services
15 Inverness Way East
Englewood, Colorado 80150
(800) 525-7052
TWX 910/935-0715

IHS offers a large number of services for users of specifications and standards. For engineers who have access to their TECHNET Search Service (described above), document ordering may be done online by way of the DOR (Documents on Request) service. Request fulfillment is either in paper copy or microform, and is processed within 24 hours.

The major IHS service is Visual Search Microfilm Files (VSMF), a collection of over 8 million pages of standards, specifications, vendor catalogs, etc. These are provided on a subscription basis in microfiche or microfilm cartridges. VSMF consists of three major data bases: the Catalog Data Base, the Industry Standards Data Base, and the Federal Government and Military Data Base. Subscribers may also select from a number of subdivisions of these, called Segmented Data Bases. VSMF files have computer-generated cross-indexing, and are updated frequently.

As noted in the descriptions of the IHS computer-search services, citations include a reference to where the full document may be located in the VSMF system.

National Standards Association (NSA)

National Standards Association
5161 River Road
Bethesda, Md. 20816
(800) 638-8076
Telex 89-8452 DISCINC BHDA

NSA also has a variety of services, many of which roughly correspond to those of IHS. NSA is the producer of Dialog's STANDARDS & SPECIFICATIONS computer-search file, described previously. Documents located in this file may be obtained directly from NSA at the above address and phone number, or may be ordered online from them by means of Dialog's DIALORDER Service.

NSA has a number of standards subscription packages available, in microfiche and some in paper copy. These include Industry Standards and Specifications, Government Standards and Specifications, and many subsets of these. All of these products are cross-indexed and updated frequently.

The newest NSA service is ASK IV, a microfiche vendor catalog information service.

Information Marketing International (IMI)

Information Marketing International
13271 Northend Street
Oak Park, Michigan 48237
(313) 546-6706
(800) 821-3031

IHS and NMA have recently been joined in this specialized field by Information Marketing International, which offers a number of microfiche and microfilm information products, involving both government and non-government documents.

Included in IMI's services are Defense Specification Service, Vendor Selector Service, Electronic Selector Service, Industry Standards Service, and a number of government publications on microfiche.

Two unusual products from IMI are:

1. A comprehensive service involving all of the environmental impact statements on microfiche—current and historical (since 1970);
2. Microfiche versions of telephone directories for over 25,000 U.S. communities and more than 2500 Canadian and international cities.

It should be noted that IHS, NSA and IMI offer other services beyond those summarized here. Anyone who is interested in any of the kinds of services mentioned here would be well advised to contact each company for complete information and latest prices.

REFERENCES

Brown, J. Standards. In *Use of Engineering Literature*. K. W. Mildren (ed.), Butterworths, London, 1976, pp. 93–114.

Hamilton, Beth A. Managing a standards collection in an engineering consulting firm. *Special Libraries*, 74:28–33, Jan. 1983.

Subramanyam, Krishna. Standards and specifications, *Scientific and Technical Information Resources*. Marcel Dekker, Inc., New York, 1981, pp. 132–148.

12

Patents

Patents have always been a major source of information for chemical and electrical engineers, but are perhaps underutilized in other engineering disciplines. Since a substantial portion of engineering designs, methods and processes are described in patent specifications, it is important that all engineers be aware of and able to use this type of information.

Nearly half a million patents are issued every year throughout the world; over 15% of these are U.S. patents. Since it is necessary that complete descriptions of the invention be included in patent applications, one can assume that almost everything that is new and original in technology can be found in patents. Yet despite the obvious wealth of information available here, patents are not extensively used by engineers as a source of technical information. In a recent article, Terapane (1976) has proposed some possible reasons for this anomaly:

- Researchers may be hindered by a lack of awareness of the kind of information which patents contain and/or they may not know how to obtain patent literature.
- The information in the patent may seem out of date as a result of the delay—often several years—between development of the invention and its acceptance and publication as a patent.
- The inefficiencies in abridgements and indices hinder certain types of searches.
- The peculiar language in which patents are written may, at first, seem difficult for researchers to comprehend.

Terapane also disproves the contention that most patent information will eventually show up in periodical literature. Over 70% of a test sample of patent technology was never published in any form of nonpatent literature.

In this chapter, we will focus on the tools and methods that will help engineers locate information about, and obtain copies of, U.S. and foreign patents. It is our intention here to provide a complement to the first volume in this series, by Konold et al. (1979) which concentrates on patent law, and to Subramanyan (1981) and other recent publications cited in the references, which provide in-depth information for the information specialist.

SOURCES OF INFORMATION ABOUT PATENTS

Printed Indexes

This section covers printed indexes that deal exclusively with patents.

Official Gazette of the United States Patent and Trademark Office. Patents. U.S. Government Printing Office, Washington, D.C. Published weekly since 1872.

The *Official Gazette* is a listing of patents issued during the week, with a brief description and sketch for each. The patents are listed in four groups: General and Mechanical, Chemical, Electrical, and Design. A companion publication, also weekly, covers trademarks.

Index of Patents Issued from the United States Patent and Trademark Office. U.S. Government Printing Office, Washington, D.C. Published annually since 1920.

This is the annual index to the *Official Gazette* weekly issues. The *Index of Patents* is published in two volumes each year—one by patentee, and the other by subject based on a classification system described below.

Manual of Classification. U.S. Patent and Trademark Office, U.S. Government Printing Office, Washington, D.C. Published looseleaf, continually revised. Also in microfiche.

In addition to the four broad subject areas used in the *Official Gazette*, the Patent and Trademark Office has an elaborate and detailed classification system for patents. The *Manual* is the printed tool for obtaining more precise subject access to U.S. Patents.

NASA Patent Abstracts Bibliography. U.S. National Aeronautics and Space Administration, Washington, D.C. Published semiannually since 1969.

Provides annotated references and indexes to NASA-owned inventions covered by U.S. patents and applications for patents that were announced in *Scientific and Technical Aerospace Reports* since May 1969.

API Abstracts/Patents. American Petroleum Institute, New York. Published weekly since 1978.

> Formerly called *API Patent Alert* and *American Petroleum Institute Abstracts of Refining Patents.*

World Patent Information. K.G. Saur Verlag, Munich. Published quarterly since 1979.

> "International Journal for Patent Information and Documentation." The text is in English. Distributed in the U.S. by Pergamon Press, Elmsford, N.Y.

Derwent Publications

Derwent Publications, Ltd., who modestly refer to themselves as "The world leaders in patents information retrieval" (Oppenheim, 1981) offer a wide variety of publications and services for accessing patent literature. Their printed publications are described briefly here, and their online and document delivery services will be covered in later sections of this chapter.

World Patents Index. Published weekly since 1974.

> This is Derwent's fast-printed, weekly current awareness service to patents issued by 26 major patent issuing authorities. It is published in four separate editions: General, Mechanical, Electrical and Chemical.

World Patents Abstracts.

> This is the generic title for a range of abstracting journals that cover eleven countries and also seven different subjects. They are all characterized by informative abstracts written in English by experts. Drawings or formulas frequently are included.

Central Patents Index. Published since 1970.

Derwent's in-depth current awareness and retrospective service for chemical patents. Separate alerting bulletins are produced for each of 12 broad headings, and a week later a *Basic Abstracts Journal* is produced containing more detailed and lengthy abstracts.

Electrical Patents Index. Published since 1980.

Similar to *Central Patents Index*. Services provided include weekly abstract bulletins, and a range of 50 subject-oriented monthly profiles.

Further information on Derwent services may be found in the article by Oppenheim (1981) cited in the references.

Other Printed Sources of Patent Information

Several report and periodical indexes include patents on a selective basis, along with other forms of literature.

Foremost among these is *Chemical Abstracts*, which has been covering, whenever possible, all chemical patents from the major technologically-developed countries since 1907. The article by Pollick (1981) describes in some detail what the Chemical Abstracts Service considers in processing patents, including patent families, equivalents, and other factors.

The reader should also note the excellent critical article by Stuart M. Kaback (1981) which compares all of the printed and computer-searchable patent indexes.

Other indexes that include some patent and patent-application citations are NASA's *Scientific and Technical Aerospace Reports* (STAR), NTIS's *Government Reports Announcements and Index* (GRA & I), and the Department of Energy's *Energy Research Abstracts*.

As noted earlier, the patent citations appearing in *STAR* are also republished separately in *NASA Patent Abstracts Bib-*

liography. NTIS has done something similar in bringing out a five-volume special publication, *Catalog of Government Patents*. This is a compendium of most patents in the U.S. Government's active patent portfolio. Vol. 1–3 cover 1966 through 1974; Vol. 4 and 5 covers 1975–1980. The basic arrangement is by subject.

NTIS also makes available special Patent Office reports, custom Patent Profile Searches, magnetic tapes, and other products and services designed to increase the use of U.S. patent technology. Along with these services, NTIS now offers a unique subscription, *Government Inventions for Licensing*. This weekly bulletin announces the more than 1,500 U.S. Government patent applications and patents issued each year.

Finally, a series of four newsletters are now on the market that should be of interest to engineers in the communications field. These are entitled:

Hi Tech Patents: Data Communications;

Hi Tech Patents: Fiber Optics Technology;

Hi Tech Patents: Laser Technology;

Hi Tech Patents: Telephony.

Each has been published twice a month since October 1982. There are also on-line versions of these newsletters, which are discussed in the next part of this chapter. For further information, the producer's address is:

Communications Publishing Group
101 Verndale St.
Brookline, Ma. 02146
617-566-2373

Computer-Search Data Bases

A variety of online data bases that focus exclusively on patents are now available. Some deal with specific subjects, some with particular geographic regions, and some try to be comprehensive. The selected listing below gives brief information, on subject, producer, online service supplier, coverage, and time span. For more complete information, including addresses and phone numbers of producers and online services, the reader should refer to:

Directory of Online Databases. Cuadra Associates, Santa Monica, Calif. Published quarterly since 1979.

APIPAT

Subject: Energy, petroleum
Producer: American Petroleum Institute
On-line Service: SDC Information Services
Coverage: 11 major industrial countries
Time Span: 1964 to date

CLAIMS/U.S. PATENTS

Subject: All areas
Producer: IFI/Plenum Data Co.
Online Service: DIALOG Information Services
Coverage: Primarily U.S.
Time Span: 1963 to date

CLAIMS/UNITERM

Subject: Chemical
Producer: IFI Plenum Data Co.
Online Service: DIALOG Information Services
Coverage: Primarily U.S.
Time Span: 1950 to date (in three files)

HI TECH PATENTS: DATA COMMUNICATIONS

Subject: Communications
Producer: Communications Publishing Group
Online Service: NewsNet, Inc.
Coverage: Full text of a newsletter of the same name. U.S., Canadian, United Kingdom Patents.
Time Span: October 1982 to date

NewsNet also offers three other full-text data bases produced by the Communications Publishing Group:

HI TECH PATENTS: FIBER OPTICS TECHNOLOGY

HI TECH PATENTS: LASER TECHNOLOGY

HI TECH PATENTS: TELEPHONY

Each is based on a newsletter of the same name, and covers the same countries as the one cited above.

INPADOC

Subject: All areas
Producer: International Patent Documentation Center
Online Service: Pergamon-Infoline
Coverage: 49 countries and the European Patent Office
Time Span: 1973 to date

A separate data base, called *INPANEW*, has recently become available from Pergamon-Infoline. It contains the most recent three weeks of patent citations to be entered in *INPADOC*.

PATDATA

Subject: All areas
Producer: BRS

Online Service: BRS
Coverage: Primarily U.S.
Time Span: 1975 to date

PATSEARCH

Subject: All areas of science & technology
Producer: Pergamon International Information Corp.
Online Service: Pergamon-InfoLine
Coverage: Primarily U.S.
Time Span: 1971 to date

USCLASS

Subject: All areas
Producer: Derwent Inc.
Online Service: SDC Information Services
Coverage: All U.S. patents by classification and number
Time Span: 1790 to date

USPATENTS

Subject: All areas
Producer: Derwent Inc.
Online Service: SDC Information Services
Coverage: Citations and abstracts for U.S. patents
Time Span: 1970 to date

WPI (WORLD PATENTS INDEX)

Subject: Chemical, electrical and mechanical patents
Producer: Derwent Inc.
Online Service: SDC Information Services
Coverage: 26 major patent-issuing authorites. Corresponds to the Derwent printed indexes.
Time Span: Varies by subject, earliest 1963

Finally, special mention should be made of CASSIS (Classification and Search Support Information System), the online search system produced by the U.S. Patent and Trademark Office. This system allows the user to determine for a particular patent number all of the classifications it is connected with, and also will display all of the patents assigned a particular classification. CASSIS is available at participating Patent Depository Libraries, which are listed later in this chapter.

SOURCES OF COPIES OF PATENTS

U.S. and foreign patents may be obtained by several methods–in person, mail, online, telephone, or telex–and from a number of sources. These may be divided into two categories: private commercial companies and not-for-profit organizations.

Private Companies

In Chapter 25 on "Libraries, Information Centers and Information Brokers," we mention the recent phenomenon of private commercial information brokers. Almost all of these include patents among the kinds of documents they can provide. Their services are speedy, but relatively expensive. Two firms that specialize in providing patent copies are:

Research Publications, Inc.
Rapid Patent Service
2221 Jefferson Davis Highway
Arlington, Va. 22202
Telephone: 703–920–5050
800–336–5010
Telex: 892362
Online thru Dialog or SDC: Order RPIPAT

This company also offers a number of other patent-connected services, including microform collections of U.S. and international patents.

Derwent Publications, Ltd.
6845 Elm St., Suite 500

McLean, Va. 22101
Telephone: 703–790–0400
Telex: 267487
Online thru SDC: Order DERWENT

Derwent is a major source of patents from 24 patent-issuing authorities, and also offers a number of indexing services and online data bases, as previously discussed.

Not-For-Profit Organizations

In this category are certain government agencies and depository libraries.

Foremost among these is the U.S. Patent Office, which can provide paper copies of U.S. and most foreign patents in-person or by mail. The address is:

Commissioner of Patents and Trademarks
Washington, D.C. 20231

Most patents and trademarks are available for 1.00 each.

United States patents are also available at a large number of public and private libraries, designated as Patent Depository Libraries. These are listed in the accompanying table. The primary method of utilizing these collections is by in-person visit.

Foreign patents may also be obtained directly from the patent-granting agency of the various countries. More information on sources of patent copies may be found in the article by Dixie L. Hunt (1982) cited in the references.

REFERENCE COLLECTIONS OF U.S. PATENTS AVAILABLE FOR PUBLIC USE IN PATENT DEPOSITORY LIBRARIES

The libraries listed herein, designated as patent depository libraries, receive current issues of U.S. Patents and maintain

State	Name of Library	Telephone Contact
Alabama	Birmingham Public Library	(205) 254–2555
Arizona	Tempe: Science Library, Arizona State University	(602) 965–7607
California	Los Angeles Public Library	(213) 626–7555 Ext. 273
	Sacramento: California State Library	(916) 322–4572
	Sunnyvale: Patent Information Clearinghouse*	(408) 738–5580
Colorado	Denver Public Library	(303) 571–2122
Delaware	Newark: University of Delaware	(302) 738–2238
Georgia	Atlanta: Price Gilbert Memorial Library, Georgia Institute of Technology	(404) 894–4508
Illinois	Chicago Public Library	(312) 269–2865
Louisiana	Baton Rouge: Troy H. Middleton Library, Louisiana State University	(504) 388–2570
Massachusetts	Boston Public Library	(617) 536–5400 Ext. 265
Michigan	Detroit Public Library	(313) 833–1450
Minnesota	Minneapolis Public Library & Information Center	(612) 372–6552
Missouri	Kansas City: Linda Hall Library	(816) 363–4600
	St. Louis Public Library	(314) 241–2288 Ext. 214, Ext. 215

State	Name of Library	Telephone Contact
Nebraska	Lincoln: University of Nebraska-Lincoln, Engineering Library	(402) 472–3411
New Hampshire	Durham: University of New Hampshire Library	(603) 862–1777
New Jersey	Newark Public Library	(201) 733–7814
New York	Albany: New York State Library	(518) 474–5125
	Buffalo and Erie County Public Library	(716) 856–7525 Ext. 267
	New York Public Library (The Research Libraries)	(212) 930–0850
North Carolina	Raleigh: D. H. Hill Library, N.C. State University	(919) 737–3280
Ohio	Cincinnati & Hamilton County, Public Library of	(513) 369–6936
	Cleveland Public Library	(216) 623–2870
	Columbus: Ohio State University Libraries	(614) 422–6286
	Toledo/Lucas County Public Library	(419) 255–7055 Ext. 212
Oklahoma	Stillwater: Oklahoma State University Library	(405) 624–6546
Pennsylvania	Philadelphia: Franklin Institute Library	(215) 448–1321†

	Pittsburgh: Carnegie Library of Pittsburgh	(412) 622-3138
	University Park: Pattee Library, Pennsylvania State University	(814) 865-4861
Rhode Island	Providence Public Library	(401) 521-7722 Ext. 226
South Carolina	Charleston: Medical University of South Carolina	(803) 792-2372
Tennessee	Memphis & Shelby County Public Library and Information Center	(901) 528-2957
Texas	Dallas Public Library	(214) 749-4176
	Houston: The Fondren Library, Rice University	(713) 527-8101 Ext. 2587
Washington	Seattle: Engineering Library, University of Washington	(206) 543-0740
Wisconsin	Madison: Kurt F. Wendt Engineering Library, University of Wisconsin	(608) 262-6845
	Milwaukee Public Library	(414) 278-3043

All of the above-listed libraries offer CASSIS (Classification And Search Support Information System), which provides direct, on-line access to Patent and Trademark Office data.

*Collection organized by subject matter.

†Call only between the hours of 10.00 a.m. and 5.00 p.m.

collections of earlier issued patents. The scope of these collections varies from library to library, ranging from patents of only recent months or years in some libraries to all or most of the patents issued since 1870, or earlier, in other libraries.

These patent collections are open to public use and each of the patent depository libraries, in addition, offers the publications of the patent classification system (e.g. The Manual of Classification, Index to the U.S. Patent Classification, Classification Definitions, etc.) and provides technical staff assistance in their use to aid the public in gaining effective access to information contained in patents. With one exception, as noted in the table above, the collections are organized in patent number sequence.

Depending upon the library, the patents may be available in microfilm, in bound volumes of paper copies, or in some combination of both. Facilities for making paper copies from either microfilm in reader-printers or from the bound volumes in paper-to-paper copies are generally provided for a fee.

Owing to variations in the scope of patent collections among the patent depository libraries and in their hours of service to the public, anyone contemplating use of the patents at a particular library is advised to contact that library, in advance, about its collection and hours, so as to avert possible inconvenience.

REFERENCES

Hunt, Dixie L. Sources of patent copies, *Science & Technology Libraries*, Vol. 2, No. 4, Summer 1982, pp. 69–78.

Kaback, Stuart M. What's new in patent information. In "Role of patents in sci-tech libraries," *Science & Technology Libraries*, Vol. 2, No. 2, Winter, 1981. (pp. 33–54)

Konold, William G. et al. *What every engineer should know about patents*. Marcel Dekker, New York, N.Y., 1979, 124 pp.

Krupp, Robert G. and Hill, Richard L. Patent and trademark collections of the New York Public Library. In "Role of patents in sci-tech libraries," *Science & Technology Libraries*, Vol. 2, No. 2, Winter, 1981. (pp. 55–60)

Meinhardt, P. Patents as a source of information for the engineer, *Use of Engineering Literature*, K.W. Mildren (ed.), Butterworths, London, England, 1976, pp. 76–92.

Oppenheim, Charles. The past, present and future of the patents services of Derwent Publications, Ltd. In "Role of patents in sci-tech libraries," *Science & Technology Libraries*, Vol. 2, No. 2, Winter, 1981. (pp. 23–31)

Pollick, Philip J. Patents and Chemical Abstracts Service. In "Role of patents in sci-tech libraries," *Science & Technology Libraries*, Vol. 2, No. 2, Winter, 1981. (pp. 3–22)

Subramanyam, Krishna. *Scientific and Technical Information Resources*. Marcel Dekker, New York, N.Y., 1981, pp. 88–99.

Terapane, John F. A unique source of information, *Chemtech*, Vol. 8, No. 5, May 1976, pp. 272–276.

ADDITIONAL READING

Special issue on patents, *IEEE Transactions on Professional Communication*, Vol. PC–22, No. 2, June 1979, pp. 46–127. Includes 17 articles on patenting, the patent process and other information.

13

Reviews and Yearbooks

Yearbooks, reviews and surveys offer engineers organized collections of information to bring or keep them up to date in a variety of fields of interest, professional society or corporate activities, or records of achievements and major events of a more general nature. Although they have the feature of currency in common, they represent assorted formats and publications.

For the engineer who must keep up to date in a subject area, annual reviews offer a wide selection of series where the engineer will find well-organized and timely surveys of various fields. Annual reviews contain references to papers, reports, articles, but more than abstracting them, they summarize their significance to the field. These reviews may be broad in scope, e.g., chemical engineering, or narrow, e.g., automatic programming. Reviews save the engineer the time of reviewing the literature and digesting and assessing major developments and trends. They afford the engineer an update or a highly accessible review of a subject when immediate information is needed for a new position or change in assignment. They are selective, organized and evaluative. Review authors must have a background in the field and expertise in identifying and evaluating the materials for the review. They are especially valuable for libraries that do not have a wide selection of books and journals.

Most of these reviews bear titles beginning *Annual Reviews . . .*, *Annual Reports . . .*, *Progress in . . .* or *Recent Developments . . .* One of the publishers supplying a broad spectrum of subjects in a review series is Annual Reviews, Inc., Palo Alto, California. This publisher has editors and editorial committees who invite qualified authors to write critical articles reviewing the important developments in major subject areas. Following are a few examples of the reviews they produce:

Annual Review of Earth and Planetary Science. Published since 1973.

Annual Review of Energy. Published since 1976.

Annual Review of Fluid Mechanics. Published since 1969.

Annual Review of Materials Science. Published since 1971.

Other series covering recent and significant developments are titled *Advances in . . .* or *Progress in . . .* or *Annual Report on . . .* The following are examples:

Advances in Chemical Engineering. Academic Press, New York. Irregular. Published since 1956.

Advances in Computers. Academic Press, New York. Irregular. Published since 1960.

Advances in Heat Transfer. Academic Press, New York. Irregular. Published since 1964.

Annual Reports on Fermentation Processes. Academic Press, New York. Published since 1977.

Progress in Materials Science. Pergamon Press, Elmsford, N.Y. Quarterly. Published since 1949.

Recent Developments in Separation Science. CRC Press, Boca Raton, Fla. Irregular. Published since 1972.

Titles such as *Advances in . . .* or *Progress in . . .* are often used for proceedings of conferences, as in *Progress in Aeronautics and Astronautics* and *Advances in Cryogenic Engineering Materials*.

As in other chapters, only an abbreviated list of titles is included. A more comprehensive list can be obtained by using this reference tool:

Irregular Serials & Annuals 1983 (8th Ed.). R.R. Bowker Co., New York, 1982, 1700 pp.

Annual reviews can be located in library card catalogs by searching under the subject for the subheadings "Collected works" or "Periodicals," e.g., Heat transfer–Collected works.

Not all reviews are packaged in these forms, however. Many state-of-the-art surveys or review articles appear in journals or reports. In contrast to review series, review articles usually cover particular research investigations. They may be detailed articles supplying information on progress in research or contain short reviews comparable to abstracting services. The service listed below is extremely helpful to engineers in locating both annual reviews and those in journals, etc.:

Index to Scientific Reviews. Institute for Scientific Information, Philadelphia, Pa. Semiannual. Published since 1974.

The *Index* has author and title word indexes. *ISR* indexes reviews in over 2900 journals and approximately 350 review series.

Yearbooks and other annuals have in common their frequency of publication, but they differ in content, in their sponsorship and their cumulation procedures. They offer general summaries, including statistics: records of accomplishments and major events of the past year in a subject area or in special research or the yearly records of societies. The label of "yearbook" is applied to an assortment of information forms. For decades, encyclopedias have been using the yearbook to update current editions. *McGraw-Hill Encyclopedia of Science and Technology* has its *McGraw-Hill Yearbook of Science and Technology*. The *Encyclopedia Americana* and the *Encyclopedia Britannica* each publish a yearbook.

Yearbooks covering transactions, activities, committees and officers of engineering and other professional societies may come to the engineer as part of a membership or a subscription to a journal or as a special issue at a separate price. Yearbooks that are essentially directories may also contain review material or statistical information. Examples of the above types follow:

Chemical Week Buyers' Guide Issue. McGraw-Hill Book Co., New York. Published since 1937.

> Received with subscription to journal. Includes raw materials, specialties, packaging, etc. Title varies.

Directory of the Transportation Research Board. National Academy of Sciences, Washington, D.C. Published since 1974.

> Information on committees and activities. Alphabetical list of individual committee members.

Human Factors Society Directory and Yearbook. The Society, Santa Monica, Calif. Published since 1959.

> Addresses and brief information about members; society officers, articles and by-laws.

Jane's All the World's Aircraft. Jane's Publishing Inc., New York. Published since 1909.

> Arranged by companies' names. Data on dimensions, weight, range, etc. of aircraft. Includes space vehicles. Photos, diagrams.

Modern Plastics Encyclopedia. McGraw-Hill Book Co., New York. Published since 1925.

> Included with subscription to *Modern Plastics*. Articles on materials, charts on properties, directory of suppliers, etc.

Oil & Gas Directory: Worldwide Exploration, Drilling, Producing Coverage. Houston. Published since 1970/71.

> Indexes companies and personnel in exploration, etc.; products.

Research in Materials: Annual Report. Massachusetts Institute of Technology, Center for Material Science and Engineering, Cambridge. Published since 1961/62.

> Summary of research for the year past.

United States Government Manual. Office of the Federal Register, National Archives & Records, General Services Administration, Washington, D.C. Published since 1934.

> Information on organization and functions of government agencies. Personnel and subject indexes.

Annual reports of companies and organizations are another form of yearbook. Although the annual reports of companies usually emphasize the financial aspects of their operation, they are helpful to engineers who may be doing business or considering employment with the company.

Almanacs are collections of miscellaneous statistics and data. Engineers will find multiple uses for their facts and figures. Probably the best known is the *World Almanac* (published by Newspaper Enterprise Association, Inc., New York).

Some journals have annual special issues covering subject areas. Representative are the following:

Annual Designer's Guide to Fluid Power Products. Hydraulics & Pneumatics. Penton/IPC Inc., Cleveland. In January issue.

> Specification charts for 45 categories of fluid power products and services; manufacturers index.

Annual Review of Extractive Metallurgy. Journal of Metals. Metallurgical Society, Warrendale, Pa. In April issue.

> Developments in physical chemistry, pyrometallurgy, lead, zinc, tin, etc.

Materials Selector. Materials Engineering. Penton/IPC Inc., Cleveland. In December issue.

> Directory of products and manufacturers. Lists of properties: metals, plastics, fibers, etc.

The classification of a publication as a yearbook is a subjective matter: That they be issued annually is the only qualification used for the items listed in this chapter.

Yearbooks can be located in library card catalogs by searching under the subject, whether it be the name of a society or a special field, for the subheading "Yearbooks," e.g., American Society for Testing and Materials–Yearbooks, or Agriculture–Yearbooks.

14

Dissertations and Theses

Doctoral dissertations are an unusual form of literature which have features that may be quite valuable to engineers in some situations. Topics for dissertations are required to be unique before they are accepted, so the work reported is frequently on the advance edge of research. They also are likely to have extensive bibliographies.

Some dissertations will be revised or rewritten for subsequent publication in periodicals or as books, but the majority will remain forever in a form that is much less accessible.

A private company, University Microfilms International (UMI), handles the announcement and distribution of most U.S. and some foreign dissertations. This company has agreements with universities with doctoral programs to prepare abstracts for all their dissertations, and to be the sole source for copies.

The universities, and their libraries, are expected to refer all requests to UMI, who currently charges nonacademic requesters $30 for a paper copy, and $17.50 for a microform. Their address, and descriptions of their major indexes, follow.

Master's theses are not given as much attention as doctoral dissertations, but some efforts are being made by UMI to announce them and make them available. Purdue University also publishes an annual index to theses in pure and applied sciences. These are also described.

INDEXES TO DISSERTATIONS AND THESES

Comprehensive Dissertation Index, 1861–1972 (37 vols.). University Microfilms International, Ann Arbor, Mich., 1973.

This major reference work lists 417,000 dissertations accepted for doctoral degrees at North American institutions. Some foreign dissertations are also included.

This publication has been kept up to date by a 19-volume, five-year cumulation (1973–1977), and five-volume annual supplements from 1978 on.

Dissertation Abstracts International. Section B. (The Sciences and Engineering.) University Microfilms International, Ann Arbor, Mich., Published monthly since 1938.

Indexes doctoral dissertations submitted for microfilming by over 450 United States and Canadian universities. Uses subject categories, such as geology and physics, for its arrangement, and includes detailed abstracts. Each issue has a keyword title index and an author index, with the latter cumulated annually.

Dissertation Abstracts International (DAI) appeared previously as *Microfilm Abstracts* (Vol. 1–11, 1938–51) and later as *Dissertation Abstracts (DA)* (Vol. 12–29,

1952–69). The current title reflects the inclusion of dissertations from a very few foreign universities. Beginning in 1966, *DA* appeared in two sections–A: The Humanities, and B: The Sciences and Engineering. The latter currently includes about 5,000 informative abstracts each year.

DAI is also available online for computer-searching, from 1861 to date, through BRS and Dialog Information Services.

Masters Abstracts. University Microfilms International, Ann Arbor, Mich., Published quarterly since 1962.

Presently includes masters theses from about 115 colleges and universities, including some foreign institutions.

Available for computer-searching from 1962 through Dialog Information Services.

Masters Theses in the Pure and Applied Sciences. Center for Information and Numerical Data Analysis and Synthesis of Purdue University, Plenum Press, New York. Published annually since 1957.

Commenced with the academic year 1955. Lists masters theses from 242 U.S. and Canadian universities. Organized by subject categories, such as Aerospace engineering and Mining and metallurgical engineering. No Abstracts.

UMI provides microform and paper copies of dissertations and theses. Orders may be placed by mail or telephone. Their address is:

University Microfilms International
P.O. Box 1764
Ann Arbor, Mich. 48106
800-521-3042

15

Trade Literature

Good trade literature supplies technical information needed by a potential user of a product or service. It includes product catalogs, specification sheets, price lists, advertisements, instruction and maintenance manuals. It can be attractively bound volumes or three-to-four-page pamphlets, charts or looseleaf pages. Engineers use trade literature in designing, developing, producing or marketing their company's products, for estimating, buying or maintenance. Engineers appreciate the value of information on components, materials, processes, equipment and systems available in their fields, since they can save development time and money by selecting parts, materials, etc. already on the market. Trade catalogs and product specifications enable them to compare the quality, performance and cost of similar products. From trade catalogs and product bulletins, engineers can obtain information not easily accessible elsewhere: application data, properties of materials, specifications, compliance with standards, performance characteristics, maintenance procedures, accessories, components and systems.

Understandably, companies and firms make it easy for engineers to acquire their information. Their advertisements in trade journals invariably include a form for obtaining literature via mail or toll-free numbers for more urgent inquiries. Addresses and phone numbers of local or regional representatives or distributors can mean instant information.

Most journals have a "New Literature," "New Products" or "Free Literature" section which offers literature with more details and data than advertisements. They usually provide readers service with a tear-out postcard for receiving further information. These "reader-reply" or "reader-request" cards need only to have appropriate numbers circled and name and address added. Many may be checked to request a phone call or visit from a salesman. Some journals are sent regularly free of charge to engineers who are potential customers for the product of their advertisers.

Besides the advertising and literature sections, many journals have a directory or buyers' guide usually free to subscribers. Some are for sale to non-subscribers, but some can be obtained only with a subscription. These consist of lists of manufacturers and product lists with means of contacting manufacturers and their representatives. The journals below are examples of those publishing such issues:

Chemical Engineering–Equipment Buyers' Guide (annual);

EDN–Product Showcase (semiannual);

Heating/Piping/Air Conditioning–HPAC Info-Dex (annual–June);

Materials Engineering–Materials Selector (annual-December);

Metal Progress—Heat Treating Buyers Guide and Directory: Equipment and Supplies, Commercial Services (annual–mid-November).

Some journals may have special issues but not on a regular basis like those above.

Directories of manufacturers and products are sometimes issued by societies, as for example, the Instrument Society of America's *ISA Directory of Instrumentation* with listings arranged in North American and international volumes.

Updating practices vary, and trade catalogs become obsolete, so it is fortunate that most are supplied free and continuously once the engineer is on a mailing list. Manufacturers expect them to pay their own way by putting them into the hands of engineers whom they hope will be recommending their products for use in their designs.

When engineers need to find the names of manufacturers or particular products, they can refer quickly to *Thomas Register of American Manufacturers and Thomas Register Catalog File* or *Sweet's Catalog File*, both described in the chapter on directories. Both supply sources for various products, but unfortunately, the catalogs they contain are by no means as numerous as the manufacturers listed in *Thomas Register*, and the rate of finding the specific product information needed at the moment or an up-to-date catalog is not high. Moreover, since compiling the *Thomas Register* or *Sweet's* is time-consuming, some of the catalog information is out-of-date before the directories are published.

The yellow pages of local telephone directories should not be overlooked. They are a somewhat limited source of product and service information, but if a manufacturer or his representative is in the vicinity, complete and current information is available immediately.

None of the above listing is a replacement for manufacturer catalogs or product specifications. However, manufacturers and service organizations are so numerous that complete files of catalogs are often out of the question for the engineer. Libraries may have token collections, but engineers or their departments usually build their own collections of practical technical information from the catalogs and other product data they use most frequently and the updates they receive once they are on a vendor's list. An article, "Building a Plant Engineering Library," which appeared in *Plant Engineering*, vol. 36, No. 2, pp. 101–120, January 21, 1982, lists sources of information on numerous processes, materials, etc. and tells how to obtain them.

More comprehensive and current sources are the compilations of trade catalogs and product specifications offered by Information Handling Services and National Standards Association. They update the technical information available from vendors and deal with the problems of filing and storage with compact units and new technology. Both use microforms, and one offers an online system which can be searched by vendor name and product or subject. Both services contain over 20,000 catalogs. Their designations and locations follow:

ASK IV Vendor Catalog Information Service (microform)
National Standards Association
Bethesda, Maryland

VSMF Visual Search Microfilm Files
Tech-Net (online i.e. computer-searchable)
Information Handling Service
Englewood, Colorado

Both of these companies also offer files of standards. These are described in the chapter on standards.

If an engineer has only a trade name, the name of the manufacturer of the product can be found by consulting the *Trade Names Dictionary*. Edited by Donna Wood, it is now in its fourth edition. Entries list trade name, product description, name of company or distributor and source of information. The *Company Index* is an alphabetical list of all companies mentioned in the *Trade Names Dictionary* and has a name and address list of over 39,000 companies. Supplements titled *New Trade Names* keep the *Dictionary* up to date.

Annual reports and company magazines can be construed as trade literature, since they usually offer information on a firm's research, development or manufacturing. Engineers seeking employment with a company will go to the interview more confidently, if they become familiar with the company's functions and products or services by reading its annual reports and/or its magazine(s). Annual reports tend to emphasize the financial aspects of companies, because they are designed to inform stockholders in the company. However, the information on other operations which they include is useful to interested engineers.

Many companies publish house organs, as they are usually called. Published for customers and potential customers, for stockholders or for employees, they print pictures and text on new developments, products or procedures. IBM's *Journal of Research and Development* and *Bell System Technical Journal* are external organs, i.e., they are designed to inform engineers and scientists outside the company. Their articles are even in some of the abstract and index services. There are also internal organs, publications designed for the company's employees and stockholders, but often distributed well beyond the company's plants and offices. They keep the employee informed and proud of the products he helps produce and create a good image and good will. *Sohio* is one of the oldest of these. House organs are listed in *Ulrich's International Periodicals Directory* (see pg. 3) under subject with the designation "House organ" at the end of the entry. They can be located by title in the index. If the company for which the engineer works does not receive a copy, the engineer can usually be added to the mailing list for the house organ in which he/she is interested in receiving.

Many companies advertise by supplying copies of reprints of articles or papers presented at conferences or meetings by their scientists and engineers. A reprint can be requested by readers using author's affiliation shown in the journal article or in the conference program. Frequently, there is a bonus of other relevant papers or reports from the author.

Engineers need technical information. Manufacturers and companies need to reach engineers. Journals, directories and special services assist both groups to communicate effectively.

REFERENCES

ISA Directory of Instrumentation. Instrument Society of America, Research Triangle Park, N.C. Annual. Published since 1979.

Sohio. Standard Oil Co. (Ohio), Cleveland. Quarterly. Published since 1929.

Trade Names Dictionary (4th Ed.). Donna Wood (ed.), Gale Research Co., Detroit, 1984, 1400 pp. (2 vols.).

Trade Names Dictionary. Company Index (4th Ed.). Donna Wood (ed.), Gale Research Co., Detroit, 1984, 1250 pp.

New Trade Names. Supplements to Trade Names Dictionary (4th Ed.), 1984, 1985.

16

Translations

Nearly half of the engineering literature in the world is written in languages other than English. American engineers have long had a poor record of learning other languages, but not even the most advanced technology can afford to ignore the scientific information produced in other countries.

A few methods are available to keep up with the foreign literature in one's special field, or to locate a translation of a particular article or book. One is to use some of the hundreds of periodicals whose contents are devoted to translated articles from foreign journals. There are two types:

1. Regular cover-to-cover translations of every issue of a particular foreign periodical;

2. Journals that contain translations of individual articles selected from various foreign sources.

COVER-TO-COVER TRANSLATIONS

This is a phenomenon dating back to the late 1940's. Periodicals chosen for such treatment are usually primary-research journals, mostly in Russian, Japanese, or German. Most are produced by commercial publishers that specialize in this area (Allerton Press, Scripta Publishing Company, Plenum Publishing Company), or scientific societies (American Institute of Physics, American Geophysical Union). Because of the great expense of such translations, the English versions are much more expensive than the foreign original; annual subscription costs of over $400 are not unusual. The engineer should also keep in mind that, because of the time required for such translating, there is a time lag between the publication of the original and the translated version, varying from six months to two years.

Several listings of cover-to-cover translations have been compiled, and are cited at the end of this chapter. A few examples of the many foreign engineering periodicals that are available in this way are cited in the following table. There are over 1000 such periodicals, so the reader is urged to consult the references cited at the end of the chapter for more complete listings.

PERIODICALS CONTAINING TRANSLATIONS FROM SEVERAL SOURCES

This type of translation journal is of even more recent vintage than the cover-to-cover titles discussed above. Most of these started in the 1960's and early 1970's. Because they are a little unorthodox, several of these titles will be described in some detail.

Title of Translation	Title of Original	Publisher
Soviet Automatic Control	Avtomatika	Scripta
Hydrotechnical Construction	Gidrotekhnicheskoe Stroitel'stvo	Plenum for ASCE
Journal of Applied Mathematics and Mechanics	Prikladnayia Mathematika i Mekhanika	Pergamon
Soviet Electrical Engineering	Elektrotekhnika	Allerton
Strength of Materials	Problemy Prochnosti	Plenum
Programming and Computer Software	Programmirovanie	Plenum
Soviet Aeronautics	Izvestiya VUZ. Aviatsionnaya Tekhnika	Allerton
Soviet Energy Technology	Energomashinostroenie	Allerton
Geotechtonics	Geotektonika	American Geophysical Union

Soviet Hydrology: Selected Papers, published by the American Geophysical Union, includes papers selected from about 30 Soviet publications in hydrology. The leading Soviet hydrology journals are scanned by the selection editor, who abstracts papers totaling 1400 to 1600 pages per year. A committee of hydrologists then selects papers totaling about 600 pages of each year's literature, which are translated into English and published. The English translation began in 1962, and is published quarterly.

International Chemical Engineering, published quarterly by the American Institute of Chemical Engineers since 1962, is subtitled, "Translations of the best in chemical engineering." Articles are selected from European and Asiatic sources. A recent issue contained 20 articles from periodicals published in Germany, France, Russia, Japan, Hungary, and Czechoslovakia.

Heat Transfer-Soviet Research is published bimonthly by Scripta Publishing Company in cooperation with the Heat Transfer Division of ASME. Each issue contains about 10 articles from various Russian sources. First published in 1969. A companion publication, *Heat Transfer-Japanese Research*, commenced in 1972.

Geodesy, Mapping and Photogrammetry is published quarterly by the American Geophysical Union. It includes the English translation of papers selected from Russian periodicals *Geodeziya i Kartografiya* and *Geodeziya i Aerofotos'yemka*. This journal, and its predecessor, go back to 1962.

About 20 periodicals of this type that are of possible interest to engineers have been identified. This listing is referenced at the end of the chapter.

TRANSLATIONS OF SINGLE ARTICLES

Several resources are available to try to locate translations of individual articles not included in the various translation journals discussed above. Examples of such material are foreign patents, articles from less prominent periodicals that haven't received the cover-to-cover treatment, articles published prior to the beginning of a cover-to-cover journal, and papers published in conference proceedings. The major sources for locating translations of specific items are described below.

Translations Register-Index. National Translations Center, John Crerar Library, Chicago. Monthly, with semiannual and annual cumulated indexes. Published since 1967.

> This publication continues *Technical Translations* (1959–1967), and *Translations Monthly* (1955–1958). The National Translations Center is a depository, referral distribution center for translations from societies, government agencies, commercial companies and many other sources. Includes patents and papers, as well as periodical articles. Translations received by NTC from 1953 through 1966 are referenced in *Consolidated Index of Translations into English*. Special Libraries Association, New York, 1969, 948 pp.

World Transindex. International Translations Center, Delft, Netherlands. Monthly, with annual cumulations. Published since 1979.

> Continues *World Index of Scientific Translations* (1967–1971) and *World Index of Scientific Translations* and *List of Translations Notified to ETC* (1972–1978). The International Translations Center was formerly called the European Translations Center. This publication covers over 30,000 translations each year, from Asiatic and East European languages into Western languages. Some translations from Western languages into French are also announced. Includes articles, patents and standards.

British Reports, Translations and Theses. British Library Lending Division, Boston Spa, England. Monthly. Published with various titles since 1971.

Formerly *BLLD Announcement Bulletin* and *NLL Announcement Bulletin.* BLLD has a very large collection of translations (430,000 as of 1979). Copies of items announced may be purchased from BLLD.

Index Translationum. Unesco, Paris, France. Annual. Published since 1950.

Lists about 6000 translations of books in the pure and applied sciences each year. Previous volumes published, 1932–1940.

In addition to these translation indexes, it is important to note that the primary technical report indexes include a substantial number of translations. Many government agencies have extensive translation activities, so sources such as *Government Reports Announcements and Index* (NTIS), *Scientific and Technical Aerospace Reports* (NASA), and *Energy Research Abstracts* (DOE) are major tools covering the government's translation report series. See the chapter on Technical Reports for further information.

It is worth noting that despite all of these indexes, the likelihood of finding a translation of a single article from a particular obscure periodical is not high, since it requires that someone else previously wanted that same article in English badly enough to pay for a translation, and further that the sponsor of the translation was thoughtful enough to provide a copy to a translation center so that it could be announced.

TRANSLATORS AND TRANSLATION SERVICES

If all of the sources and indexes cited above have failed, and the need is great enough, it may be necessary to commission a translation of the original foreign-language article.

Three good sources for locating individuals and firms that specialize in the required language and discipline are cited below.

Translation & Translators: An International Directory and Guide. Stefan Congrat-Butlar, R. R. Bowker, New York, 1979, 241 pp.

Translators and interpreters are arranged in two categories: literary and industrial, and scientific and technical. The index is by language.

Translator Referral/Translation Services Directory. Translation Research Institute, Philadelphia, Pa. Annual. Published since 1974.

The Translation Research Institute was formerly known as the Guild of Professional Translators.

Literary Market Place, With Names and Numbers: The Directory of American Book Publishing. R. R. Bowker, New York, 1982.

Section 34 (pp. 360–368) of the 1982 edition is entitled, "Translators." Individuals and organizations are listed alphabetically. Entries include indication of language and subject specialties. A classified language index is appended.

REFERENCES

Listings of Cover-To-Cover Translations

Selected list of translated journals. In *Scientific and Technical Information Resources*. Krishna Subramanyam, Marcel Dekker, Inc., New York, 1981, pp. 274–281.

A Guide to Scientific and Technical Journals in Translations (2nd ed.). Carl J. Himmelsbach and Grace E. Brociner, Special Libraries Association, New York, 1972, 49 pp.

Cover-to-cover translations of potential interest to engineers. In *Use of Engineering Literature*. K. W. Mildren (ed.), Butterworths, London, 1976, pp. 55–60.

Journals in Translation. British Library Lending Division and International Translations Centre, BLLD, Boston Spa, England, 1978, 181 pp.

A 1982 edition of this publication has been announced.

Listings of Periodicals Containing Selected Translations

Journals consisting of translations of articles selected from various foreign journals. In *Use of Engineering Literature*. K. W. Mildren (ed.), Butterworths, London, 1976, pp. 60–61.

ADDITIONAL READING

Role of translations in sci-tech libraries. *Science & Technology Libraries*, Vol. 3, No. 2, Winter 1982.

A special issue containing six articles on translations:

a. Printed and online sources for technical translations. Suzanne Fedunok (pp. 3–12);

b. The National Translations Center: Its development, scope of operation and plans for the future. Ildiko D. Nowak (pp. 13–19);

c. The role of commercial translation firms in providing

technical material to sci-tech libraries. Robert L. Draper (pp. 21–29);

d. The translator in the United States. Fred Klein (pp. 31–46);

e. CUTTA serves: Organizing translation within a university. Victor Hertz (pp. 47–55);

f. Ei's inside look at technical translation. Zoran Nedic, Barbara S. McCoy (pp. 57–63).

17

Conferences

Presenting a paper at a scientific meeting is one of the most popular methods of announcing the results of research. The number of meetings and conferences has increased along with the amount of research being performed. According to a recent estimate, over 10,000 conferences are held each year all over the world, and the new *Engineering Index* data base, *Engineering Meetings*, covers over 2000 conferences annually.

Many conferences do not publish proceedings, but give the authors the right to publish their papers in journals or other publications at a later date. Many of these papers will never be published, or published as originally presented at the meeting. The author may be the only source for a copy of the paper. Some conferences publish only abstracts of papers, and others publish the proceedings in scientific journals.

Various bibliographic aids have been created to provide access to current and retrospective records of these meetings, and the availability of publications resulting from these meetings. These tools fall naturally into two categories: Those providing information on forthcoming meetings; and those giving details of the publications arising from conferences.

FORTHCOMING CONFERENCES

One of the most common methods that engineers use to learn about upcoming meetings is to scan the news or calendar sections of periodicals. Following are some examples of periodicals that regularly include announcements of upcoming meetings: *Mechanical Engineering; Aviation Week and Space Technology; Industrial Engineering; Electronic Design; Civil Engineering; Environmental Science and Technology; IEEE Micro;* and *IEEE Spectrum*.

In addition, there are a number of more formal information resources that are specifically devoted to announcing forthcoming conferences. Each of these are described below.

Scientific Meetings. Scientific Meetings Publications, San Diego. Published quarterly since 1957.

> Describes future meetings of technical, scientific, medical, and managerial organizations and universities.

World Meetings: United States and Canada. Macmillan, New York. Published quarterly since 1963.

> One of a series of publications announcing forthcoming meetings. The one other of possible interest to engineers is *World Meetings: Outside United States and Canada.*

Forthcoming International Scientific and Technical Conferences. Aslib, London. Published quarterly since 1923?.

The main issue of each year is published in February. Supplements are published in May, August, and November.

Meetings on Atomic Energy. International Atomic Energy Agency, Vienna. Published quarterly since 1969.

Distribution of this publication in the U.S. is handled by Unipub, New York.

World Calendar of Forthcoming Meetings: Metallurgical and Related Fields. The Metals Society, London, Eng. Published quarterly since 1965.

PAPERS & PUBLICATIONS RESULTING FROM MEETINGS

The papers presented at conferences may eventually appear in a number of ways:

- As an article in a journal;
- As a technical report;
- In a proceedings volume.

Papers that are published in periodicals can be located by means of *Engineering Index*, and the various other periodical indexes discussed in the chapter on *Abstracts and Indexes*. The technical report indexes, such as *Government Reports Announcements and Index, Scientific and Technical Aerospace Reports*, and *Energy Research Abstracts*, cover papers and conference publications in report format very thoroughly.

For proceedings volumes, there are several printed and computer-search indexes that provide information on proceedings publications and papers. Most of them list the proceedings in a variety of ways, including keyword, sponsor, date, editor and location.

Directory of Published Proceedings. Series SEMT: Science/Engineering/Medicine/Technology. InterDok Corporation, Harrison, N.Y.

A bibliographic directory of preprint volumes and published proceedings. Arranged chronologically, with keyword indexes for name of conference, sponsors and title.

InterDok also produces another series of this directory that may be of interest to engineers, *Series PCE: Pollution Control & Ecology*. InterDok provides an order service for obtaining the proceedings included in the directory.

Proceedings in Print. Proceedings in Print, Inc., Arlington, Mass. Published bimonthly since 1964.

Originally restricted to aerospace technology, *PIP* now covers proceedings in all disciplines and all languages. Indexes about 2000 items a year.

Irregular Serials & Annuals; An International Directory. R. R. Bowker, New York. Published every two years since 1970.

Similar in arrangement to *Ulrich's International Periodicals Directory*, the 8th edition (1982) of this publication provides data on some 37,000 serials, annuals, continuations, conference proceedings and other publications issued irregularly or less frequently than twice a year.

Yearbook of International Congress Proceedings. Union of International Associations, Brussels. Published annually since 1969.

Provides coverage of proceedings, reports, symposia, and

other documents emanating from international congresses.

Index to Scientific & Technical Proceedings. Institute for Scientific Information, Philadelphia. Published monthly since 1978.

One of the very few tools available that indexes individual papers included in proceedings volumes. Covers over 90,000 papers from 3,000 published proceedings, including those appearing as periodicals, multiauthored books, and report series.

This Index is available online for computer-searching, but only from the publisher through their own search service.

Conference Papers Index. Cambridge Scientific Abstracts, Bethesda, Md. Published monthly since 1973.

Formerly entitled *Current Programs* from 1963 to 1972. Contains listings of papers presented at conferences in the fields of life sciences, medicine, engineering and technology, chemistry, and the physical sciences.

Conference Papers Index is available for computer-searching from Dialog Information Service.

The most recent addition to the specialized tools available in this area is *Ei Engineering Meetings*, a service that is accessible only by computer-searching.

The Ei Engineering Meetings database covers significant published proceedings of engineering and technical conferences, symposia, meetings, and colloquia. Each meeting included is indexed in a main conference record. In addition, all papers from the meeting are individually indexed. Records in the file do not contain abstracts. It offers complete coverage of meetings held since July 1982, and selective coverage of meetings from 1979 to June 1982. Approximately 2000 published proceedings are indexed each year.

The database is produced by Engineering Information, Inc., who also publish *Engineering Index*, and is available through the Dialog and SDC computer-search services.

Finally, as noted earlier, the author may be the best (and perhaps the only) source for a copy of a paper presented at a conference. A card or letter asking for a preprint will sometimes succeed when all of the traditional information resources have failed.

ADDITIONAL READING

Conference Literature: Its Role In The Distribution of Information. Gloria J. Zamora and Martha C. Adamson (eds.), Learned Information, Marlton, N.J., 1981, 240 pp.

18

Professional Societies

Professional organizations, societies, and associations provide a variety of valuable services for practicing engineers. Some of the more obvious are preparation of standards, publication of books and periodicals, and the sponsorship of technical conferences. Additional services offered by some societies include maintenance of libraries, placement services, and speaker's bureaus; awards and scholarships; and special cooperative programs for students and industry.

In some areas, there are engineers' clubs that, although not specifically affiliated with any national organization, provide many of these same kinds of services. Also, many of the larger engineering associations have local or regional chapters which provide opportunities for participation to engineers who are unable to attend national conferences.

Most nation-wide societies, and many of the regional chapters, publish membership directories usually giving affiliations and telephone numbers. These can be very helpful when trying to contact a colleague who doesn't appear in the biographical directories discussed in the *Directories* chapter. Distribution of these directories is normally to members, but they are often available to others for a nominal charge.

Several bibliographic tools are available to provide information about professional societies, their programs and publications. Some of these have also been discussed briefly in other chapters.

Encyclopedia of Associations. Gale Research Co., Detroit. Published since 1956, and annually since 1975.

> "A guide to national and international organizations, including: trade, business, and commercial; agricultural and commodity; legal, governmental, public administration and military; scientific, engineering and technical; educational; cultural; social welfare; health and medical; public affairs; fraternal, foreign interest, nationality and ethnic; religious; veteran, hereditary and patriotic; hobby and avocational; athletic and sports; labor unions, associations and federations; chambers of commerce; and greek letter and related organizations."

The 17th edition, 1983, is divided into three volumes:

Vol. 1: National Organizations of the U.S.;

Vol. 2: Geographic and Executive Indexes;

Vol. 3: New Associations and Projects (Inter-edition Supplement to Volume 1).

Section 4 of Volume 1 is entitled: "Scientific Engineering and Technical Organizations." It includes 1171 associations in the following areas: aerospace, anthropology, architecture, astronomy, behavioral sciences, biology, botany, chemistry, demography, ecology, electronics, energy, environmental quality, genetics, geology, information processing, meteorology, nuclear physics, oceanography, paleontology, parapsychology, phenomena, psychology, standards, water resources.

A typical entry is reproduced below:

★5023★ INSTITUTE OF ELECTRICAL AND **ELECTRONICS** ENGINEERS (IEEE)
345 E. 47th St. Phone: (212) 705-7900
New York, NY 10017 Eric Herz, Exec.Dir.
Founded: 1963. **Members:** 210,000. **Staff:** 350. Regions: 10. Sections: 242. Subsections: 44. **Local Groups:** 530. Engineers and scientists in electrical engineering, electronics, and allied fields; membership includes 30,000 students. Holds numerous meetings and special technical conferences. Conducts lecture courses at the local level on topics of current engineering and scientific interest. Assists student groups. Awards medals, prizes, and scholarships for outstanding technical achievement. Supports Engineering Societies Library in New York City with other groups. Boards: Awards; Educational Activities; Publications; Regional Activities; Technical Activities; U.S. Activities Board. **Councils:** Oceanic Engineering; Solid-State Circuits. Societies: Acoustics, Speech and Signal Processing; Aerospace and Electronic Systems; Antennas and Propagation; Broadcast, Cable and Consumer Electronics; Circuits and Systems; Communications; Components, Hybrids and Manufacturing Technology; Computer; Control Systems; Education; Electrical Insulation; Electromagnetic Compatibility; Electron Devices; Engineering Management; Engineering in Medicine and Biology; Geoscience Industrial Electronics; Industry Applications; Information Theory; Instrumentation and Measurement; Magnetics; Microwave Theory and Techniques; Nuclear and Plasma Sciences; Power Engineering; Professional Communications; Quantum Electronics and Applications; Reliability; Social Implications of Technoloy; Sonics and Ultrasonics; Systems, Man and Cybernetics; Vehicular Technology. **Publications:** (1) Proceedings, monthly; (2) Spectrum, monthly; (3) Directory, annual; (4) Standards, irregular; the Societies and Councils publish 42 journals, 8 magazines, and 125 conference proceedings. **Formed by Merger of:** American Institute of Electrical Engineers (founded 1884) and Institute of Radio Engineers (founded 1912). **Convention/Meeting:** irregular technical shows, conferences and symposiums.

It is important to note that although the title of Volume 1 is "National Organizations of the U.S.," it includes international associations if they have a substantial number of U.S. members.

Scientific, Engineering, and Medical Societies Publications in Print, 1980–1981. (4th ed.), James M. Kyed and James M. Matarazzo (Compilers), R. R. Bowker Co., New York, 1981, 626 pp.

This volume provides a listing of print and non-print material produced by and presently available from professional societies. Included are monographs, journals and periodicals, conference proceedings, specifications, standards, brochures, pamphlets, manuals, tapes, maps, slides, and charts. The material is arranged in alphabetical order by society. Each entry gives the address, phone number, and ordering information, as well as complete bibliographic information.

This edition will probably be the last. Bowker has now begun publishing a more comprehensive reference book which effectively supercedes this one. See the next item.

Associations' Publications in Print. R. R. Bowker, New York. Published annually since 1981.

This two-volume set lists publications of nearly 3000 associations and societies. Vol. 1 is a subject index, while Vol. 2 contains indexes for title, publisher/title, association name, and acronym. The first edition (1981) included more than 71,000 publications.

Special mention should be made of the American Association of Engineering Societies. AAES is a national organization with 43 member organizations representing nearly one million engineers. As a central coordination organization, AAES focuses the resources of the engineering societies on technological issues of national and international importance. Two of their publications are described below.

Directory of Engineering Societies & Related Organizations. American Association of Engineering Societies, New York. Published every two or three years since 1956.

The latest edition is the 10th, dated 1982. This publication lists nearly 500 national, regional, Canadian, and international organizations having primarily engineering, scientific, or related activities. Each society, association, club, or fraternity is listed by name with its official abbreviation, address, and telephone number. Also shown are the name and title of the principal officer and all elected officers; staff size; total membership; founding date; budget; membership requirements; and objectives.

Who's Who in Engineering. American Association of Engineering Societies, New York. Published since 1970.

The latest edition is the 5th, dated 1982. The first and second editions were titled: *Engineers of Distinction.* The major part of the book contains biographies of engineers, but the first 56 pages provide information on awards given by various engineering societies that is not included in the directories cited above. Included for each society is a listing of its past presidents, names of the awards it gives, a statement of the purpose of the award, and a list of all recipients.

As noted above, one of the functions of professional societies is the sponsorship of meetings. The papers and proceedings they generate are covered in Chapter 17 on "Conferences."

19

Maps and Atlases

Engineers use special types of maps and charts in a variety of applications in the practice of their profession. Many an engineering project calls for the use of maps and charts, so it is essential that engineers know the types available and how and where to locate them.

Probably the heaviest users are civil engineers who need topographic maps to study the surface features of the terrain in regions where bridges or higways are being planned, for example. Topographic maps show location, shape and size and elevation of hills, bodies of water and other physical features by means of contour lines, color or shading or relief representations. The U.S. Geological Survey prepares topographic maps of the states and offers them for sale. Maps of areas east of the Mississippi can be purchased at the Eastern Distribution Branch of the Geological Survey in Arlington, Virginia; those of areas west of the Mississippi at the Western Distribution Branch in Denver. All Geological Survey maps can be purchased over the counter at U.S. Geological Survey offices in several U.S. cities including Denver, Los Angeles and Reston, Virginia. The Survey's products can also be ordered from the National Cartographic Information Center, described below. The topographic maps have a well-developed indexing system which should be consulted to locate topographic maps, especially those of urban areas. These may not bear the name of the city, but rather the name of some other geographic feature.

Another important source is the Defense Mapping Agency which produces and distributes maps, charts including nautical charts and marine navigational data.

Thematic maps, often referred to as topical, distribution, statistical or special maps, illustrate a theme or subject, for example, soil, precipitation or mineral resources. Petroleum engineers might consult bathymetric maps for ocean depths. Mechanical engineers might use economic maps to determine the manufacturing in a geographical area. Industrial engineers might use demographic maps for population density near a potential plant site. Chemical engineers are becoming increasingly responsible for the disposal of chemical wastes and must know soil composition, location of water supplies and routes for transporting waste. A look at the following list of a few of the existing types of thematic maps will suggest their relevance and other applications in the various fields of engineering: geologic, gravity, hydrological, land-use, planning, political/administrative, tectonic, vegetation.

Especially useful to civil engineers is *Guide to USGS Geologic and Hydrologic Maps* which lists geologic and hydrologic maps, including oil and gas investigations charts and maps, coal investigation maps and mineral investigations resource maps. *Handbook of Geology in Civil Engineering* devotes an entire chapter to geological maps.

Charts are a map form. Nautical charts, also known as hydrographic or marine charts, and lake charts are extensively used for navigation. Charting is useful in developing hydroelectric or irrigation plans, flood control or fisheries. Governments are the largest producers of maps and charts. Charts of U.S. waters are prepared by the U.S. Geological Survey; those of foreign waters by the U.S. Naval Oceanographic Office.

Maps can appear in sheet form, in books or periodicals, reports or proceedings of professional societies. Atlases are bound collections of maps, charts, tables or plates illustrating a variety of subjects. Geographical atlases cover a vast territory—any place from individual states of the U.S. to planets. There are also atlases of minerals, galaxies and various parts of the human anatomy. Atlases can be searched in library card catalogs under the term, Atlases.

Many libraries have extensive map collections. A directory of over 600 of these, *Map Collections in the U.S. and Canada*, describes the types and subjects collected. Libraries in all parts of the U.S. have been selected as depositories for government-produced maps and contain those of the USGS and, often, those of the Defense Mapping Agency.

There are numerous map sellers. The closest at hand can be located in the yellow pages of local telephone directories. Books on maps often list map publishers and sellers and include such companies as American Map in New York, Denoyer-Geppert in Chicago and Hammond in Maplewood, N.J. Most publishers and sellers issue catalogs. Important sources for maps and atlases are publishers' catalogs, including those of foreign and domestic government agencies and commercial map publishers; and geological and other societies. A good source of information on new types and themes is *Bibliographic Guide to Maps and Atlases*, an annual subject list of publications added to the Research Libraries of the New York Public Library and the Library of Congress. All languages and book and non-book materials are included.

There are many pertinent books on the use of maps covering such points as scale and other details of computation. One example is *Understanding Maps*. For engineers who must prepare and produce maps there is *Practical Map Production*. Other books can be found in the subject sections of library card catalogs under the heading, Maps.

Gazetteers list alphabetically names of mountains, rivers and other geographic features and give brief identifying information about them. They are usually the first step in choosing the correct maps for locating a town, a lake, etc. because they provide the geographic or position coordinates and the political or administrative unit in which the place is located. The Times' *Index-Gazetteer of the World* and the *International Geographic Encyclopedia and Atlas*, fulfill this function. Other gazetteers can be located by searching the card catalog under Gazetteers, or Geography—Dictionaries.

Human resources—map librarians and map curators of societies and agencies issuing maps—offer engineers expertise in using all of the above information resources effectively and suggesting others. The staff of the National Cartographic Information Center, U.S. Geological Survey, will fill requests for information at the Center in Reston by mail or phone (703/860–6045). They can advise requestors about the maps and other cartographic information that can be obtained from many government and private sources and search its holdings to satisfy a request for specific information.

REFERENCES

Bibliographic Guide to Maps and Atlases. G. K. Hall, Boston. Annual. Published since 1979.

Commercial Atlas & Marketing Guide. Rand McNally & Co., Chicago. Annual. Published since 1870.

Guide to USGS Geologic and Hydrologic Maps. Documents Index, McLean, Va., 1983, 644 pp.

Handbook of Geology in Civil Engineering (3rd Ed.). Robert F. Legget and Paul F. Karrow, McGraw-Hill Book Co., New York, 1983, 1308 pp.

Index-Gazetteer of the World. Times Publishing Co., London, 1965, 964 pp.

International Geographic Encyclopedia and Atlas. Houghton Mifflin Co., Boston, 1979, 890, A1–115 pp.

International Maps and Atlases in Print (2nd Ed.). Kenneth L. Winch (ed.), R. R. Bowker Company, New York, 1976, 866 pp.

Map Collections in the United States and Canada (3rd Ed.). David K. Carrington and Richard W. Stephenson (eds.), Special Libraries Association, New York, 1978, 230 pp.

National Atlas of the United States. U.S. Geological Survey, Washington, D.C., 1970, 417 pp.

National Gazetteer of the United States of America (Geological Survey Professional Paper 1200). Government Printing Office, Washington, D.C., 1982– (Series begun in 1982 will list geographic names for places, features and areas in volumes for each state and territory. Includes quadrants on topographic USGS maps.)

National Geographic Atlas of the World (5th Ed.). National Geographic Society, Washington, D.C., 1981, 383 pp.

Practical Map Production. John Loxton, John Wiley & Sons, New York, 1980, 137 pp.

Road Atlas: United States, Canada, Mexico. Rand McNally & Co., Chicago. Annual. Published since 1932.

Understanding Maps. J. S. Keates, Longman, Inc., New York, 1982, 139 pp.

World Directory of Map Collections. Walter W. Ristow (ed.), Verlag Dockumenation, Munich (for the International Federation of Library Associations), 1976, 326 pp.

20

Statistical Information Sources

One of the most important contributions of statistics to engineering is support in problem-solving and decision-making. Engineers use statistics to plan, to report, to convince. Typical are the engineer selecting a plant site who must know population figures; the engineer estimating cost of materials for a proposal; the engineer who must justify manpower and equipment budgets. When statistics are required, engineers must be aware of sources such as books, journal articles and reports that supply them. The following sources can be used to locate statistics on a variety of subjects:

American Statistics Index: A Comprehensive Guide and Index to the Statistical Publications of the United States Government.

Congressional Information Service, Inc., Bethesda, Md. Monthly. Published since 1973.

> Issues in two parts: Abstracts arranged by originating agency; index arranged by subject.

Directory of International Statistics (Vols. 1–2). United Nations, New York, 1982–

> Vol. 1:Pt. 1. Listing of international statistical series including trade, environment and transportation statistics; Pt. 2. Banks of economic and social statistics (in preparation).

Statistical Reference Index. Congressional Information Service, Inc., Bethesda, Md. Monthly. Published since 1980.

> Selective guide to American statistical publications from sources other than the U.S. government. Access to sources of economic, social and demographic data in the publications of business and research organizations, universities, associations et al.

Statistical Sources (17th Ed.). Paul Wasserman and Jacqueline O'Brien, Gale Research Co., Detroit, 1982, 975 pp.

> Sources of current statistical data arranged by subject: industrial, business, social, educational, financial and other topics for the U.S. and the world.

Statistics appear frequently in periodicals, for example, price information or consumption of various products, and can be located through indexing services such as *Applied Science & Technology Index* or *Monthly Catalog of United States Government Publications*. The subheading "Statistics" should be searched under the appropriate subject.

Government agencies at both the national and international level are the source of a vast amount of statistical data, but statistical information is also available from trade and professional organizations, commercial publishers and companies and universities. Statistics appear in books, journals and government documents. The following list illustrates the variety of their sources:

Aluminum Statistical Review. Aluminum Association, New York. Annual. Published since 1962.

Engineering Manpower Bulletin. American Association of Engineering Societies, Irregular. Published since 1976.

ENR (Engineering News-Record) Quarterly Cost Roundup. McGraw-Hill, Inc. Published in appropriate issues of the weekly magazine.

Information on construction materials, wages, etc.

Fact and Figures for the Chemical Industry. Chemical and Engineering News. American Chemical Society, Washington, D.C. Annual. Published since 1956.

Figures on production, prices, employment.

Minerals Yearbook. U.S. Bureau of Mines, Washington, D.C. Annual. Published since 1932.

Data on metals, minerals, fuels; mineral industries of the states.

Quarterly Oil Statistics. Organisation for Economic Co-operation and Development, Paris. Published since 1974.

Crude oil, oil products and natural gas supplies and disposition information arranged by country. Import/export data.

Science Indicators. U.S. National Science Foundation, National Science Board, Washington, D.C. Biennial. Published since 1973.

Quantitative assessment of science and technology with extensive statistical tables on R&D expenditures, employment of scientists and engineers, patents, etc.

Statistical Abstract of the United States. U.S. Bureau of the Census, Washington, D.C. Annual. Published since 1978.

Quantitative summary statistics on political, social and economic organization of the U.S. Supplies information on sources of all tables.

Statistical Yearbook. United Nations, New York. Published since 1948.

Summary of international statistics on mining, manufacture, trade, etc. of various countries.

Transportation Energy Data Book (6th Ed.). G. Kulp and M. C. Holcomb, Noyes Data Corp., Park Ridge, N.J., 1982, 175 pp.

Statistics characterizing transportation activity and factors influencing transportation energy use. Sections cover highway, air, water, etc. (DOE report ORNL–5883)

World Demand for Raw Materials in 1985 and 2000. Wilfred Malenbaum, E/MJ Mining Information Services, New York, 1978, 126 pp.

Estimates changes in level of demand for materials and distribution among geographic regions.

Compilations of statistics can be located in library card catalogs by searching the terms "Statistics" and "Statistical Methods" as subheadings under specific subjects such as Engineering, Research, Agriculture, etc.

21
Audio-Visual Materials

Although audio-visual (A-V) materials are an ideal medium for the continuing education of engineers, surprisingly few are publicized. However, with academic institutions, societies and commercial firms realizing the potential of featuring authorities in their fields presenting lectures, short courses and demonstrations, it is reasonable to assume that there will be increased activity in production. These might be used at a plant by a group of engineers seeking information about a concept or materials in preparation for a new project. They might be used for a program at a professional meeting. If they replace the cost of travel or fees for speakers, their rental fees or purchase prices would be considered reasonable.

Two videotapes produced by the Massachusetts Institute of Technology serve as examples of the potential of A-V. The brochures for "Digital Signal Processing," featuring Alan V. Oppenheim, and "The Deming Videotapes: Quality, Productivity, and Competitive Position," featuring W. Edwards Deming, describe them as appropriate "for the practicing engineer, scientist and technical manager."

Films and other audio-visual materials that are available can be located in directories such as the *Educational Film Locator*. . . or the NICEM film indexes. Both have subject indexes. Large academic and public libraries will have the directories noted above and further information on A-V materials.

REFERENCES

Educational Film Locator of the Consortium of University Film Centers and R. R. Bowker Company (2nd Ed.). R. R. Bowker Co., New York, 1980, 2611 pp.

Audiovisual Market Place: A Multimedia Guide. R. R. Bowker Co., New York. Annual. Published since 1969.

Lists producers, distributors and services under types of films, etc. i.e., industrial, scientific, etc.

Science Books & Films. American Association for the Advancement of Science, Washington, D.C. Five issues per year. Published since 1965.

"Critical reviews of books, films and filmstrips in mathematics and the social, physical and life sciences". Reviews are coded with level of content, i.e., Professional, General Audience, etc.

Since 1964 the National Information Center for Educational Media (NICEM), University of Southern California, Los Angeles, has provided indexes and online search service information for various media. Several contain items of interest to engineers:

Index to 16 mm Educational Films (7th Ed.) (Vols. 1-4);

Index to Educational Video Tapes (5th Ed.);

Index to Environmental Studies–Multimedia (2nd Ed.);

Index to Health and Safety Education–Multimedia (4th Ed.).

The indexes are updated and new editions issued regularly. The NICEM database is available via DIALOG (file 46).

22

Engineering Software

Several chapters in this book deal with computer-searching as a means of accessing journal articles, technical reports, standards and specifications, and patents. These searches utilize databases prepared by government or commercial agencies and are usually performed by librarians or information specialists who can keep current with new bases and searching techniques. This chapter will deal with the sources of the software which engineers will use or adapt for the microcomputers which are changing the nature and distribution of their work.

Software is considered here simply as the programs needed to make computers perform their prescribed functions or calculations, including operating systems, utility programs and applications packages. The term "software" is used to denote all types of programs usually those whose use is not limited to one specific job or application.

Computers are becoming increasingly portable and inexpensive, but if engineers do not wish to become programmers, they must be aware of the software available and compatible with their hardware. If suitable or adaptable software for their applications has already been created, the engineers can save time by using the programs already developed. Software is often the critical factor in choosing hardware.

These sources of software help users locate appropriate packages: books, regularly appearing sections in journals, software search services, databases and information centers. Examples follow:

BOOKS

BASIC Programs in Production and Operations Management. Pricha Pantumsinchai et al, Prentice-Hall, Inc., Englewood Cliffs, N.J., 1983, 443 pp.

> BASIC, command-driven interactive programs for solving production and operations management problems.

BASIC Reservoir and Drilling Engineering Manual. John L. Crammer, Jr., PennWell Books, Tulsa, Okla., 1982, 232 pp.

> All programs in this book available on diskettes for input on a TRS 80 Model II.

Computer Programs for Electronic Analysis and Design. Dimitri Bugnolo, Prentice-Hall, Inc., Englewood Cliffs, N.J., 1983, 288 pp.

> Collection of computer programs for analysis and design of solid-state electronic circuits.

IIE Software. Institute of Industrial Engineers, Norcross, Ga., 1983.

A series of TRS–80, Apple and IBM-PC software. Program groups include production control, work measurement, forecasting, project management and economic analysis.

Science and Engineering Programs for the IBM-PC. Cass R. Lewart, Prentice-Hall, Inc., Englewood Cliffs, N.J., 1983, 150 pp.

Twenty programs in electronic engineering, number theory, computer program design, data communications, probability, statistics, operations research and applied mathematics.

PERIODICALS

Advances in Engineering Software. C.M.L. Publications in association with the International Society for Computational Methods in Engineering, Ashurst, Southampton. Quarterly. Published since 1978.

Stated aim of the journal is to act as a "link between the originators of software and the engineering community . . ." Subjects of articles include theory and scope of applications.

Computers in Mechanical Engineering (CIME). American Society of Mechanical Engineers, New York. Bimonthly. Published since 1982.

Section titled "Software Exchange" has items divided under "Microcomputers" and "Mainframes and Minis." Describes capability and gives information on computer language, hardware, developer of the software and its availability and cost. Directions for submitting programs to the Exchange are included in the section each issue.

Engineering Software Exchange. Yonkers, N.Y. Bimonthly. Published since 1983.

Informs engineers of new high-level technical software and offers a free service for individuals who wish to sell or exchange technical software for microcomputers.

IEEE Micro. IEEE Computer Society, Los Alamitos, Calif. Bimonthly. Published since 1981.

IEEE Micro's "Product Summary" is a regular feature of each issue and contains a section on software.

DIRECTORIES AND DATABASES

Catalog of Directories of Computer Software Applications. National Technical Information Service (NTIS), Springfield, Va., 1983, 12 pp.

Contains brief descriptions of directories available on 25 subjects including aerodynamics, civil and structural engineering, electrical and electronics engineering and environmental pollution and control. All are available from NTIS.

Directory of Computer Software, 1983. Walter L. Finch, National Technical Information Service, Springfield, Va., May 1983, 200 pp. (NTIS order # PB83–167668)

Unique guide to machine-readable software compiled from more than 100 U.S. government agencies. Describes more than 500 programs with coverage in subject categories such as civil and structural engineering, computer sciences, mathematics and statistics. Programs for application software, graphics software, software tools and modeling and simulation programs.

Major Software Sources for the Consulting Engineer. American Consulting Engineers Council, Washington, D.C., 1983, 140 pp.

> More than 300 programs in fifteen categories indexed by software vendors.

International Software Database. Imprint Software Ltd., Fort Collins, Colo. (DIALOG file 232) from 1973 to the present. Monthly updates.

> Over 10,000 records in 1983. Comprehensive list of commercially available software for any type of mini- or microcomputer. Brief description of software items with coding for compatible equipment. Purchase information and order service available.

Software Catalog: Science and Engineering. Elsevier Science Publishing Co., New York, 1984, 525 pp.

> Created from *International Software Database.* Information on availability, cost, applications and compatibility of over 4,000 existing micro- and minicomputer packages and the systems on which they run. Appropriate software can be located by programming language, microprocessor, applications, computer or operating system or keyword.

INFORMATION AND SERVICE CENTERS

Data & Analysis Center for Software (DACS), Rome Air Development Center, ISISI, Griffiss AFB, N.Y. 13441

> Subject coverage includes modern programming practices, standards and guides for software development and maintenance techniques and tools. Maintains a database on software development and maintenance pro-

grams. The *DACS Annotated Bibliography* is a printed copy version of the information contained in its Software Engineering Bibliographic (SEB) Database (available NTIS).

National Software Exchange. 38 Melrose Pl., Montclair, J.J. 07042.

Arranges for members to trade used software for an annual fee and handling charge. These are original programs accompanied by all documentation.

U.S. Department of Energy National Energy Software Center, Argonne National Laboratory, 9700 S. Cass Ave., Argonne, Il 60439.

Has a computer software library and software information center services. Subject coverage includes chemistry and chemical engineering, computers, data processing, systems, electronics and electrical engineering, energy, environmental engineering, mathematics, mechanical engineering, metallurgy, mining, nuclear science, pollution, physics.

Micro Associates, Inc., 2300 Highway 365, LB131, Nederland, Tex. 77627

Packages for a variety of micro/personal computers for design of vacuum and pressure vessels and flanges.

Practical Engineering Applications Software, 7208 Grand Ave., Neville Island, Pittsburgh, Pa. 15225.

Ready-to-run programs include packages for area and rotating moments of inertia, column and spring design, sheave and sprocket design, heat transfer, and project management.

Stress Analysis Associates, 285 North Hill Ave. #201, Pasadena, Calif. 91106.

> Lists of programs written in BASIC. Capabilities such as static analysis, analysis of beams, calculation of stresses in disks and plates, heat transfer, hydrodynamics.

Information on addresses, phone numbers and brief notes on packages for the last three and other companies is contained in the following article:

Falk, Howard. (1983). Software tools for mechanical engineers. *Mechanical Engineering* 105:28–38.

Several hundred programs are ready to provide computer-aided design modelling and simulation capabilities in most types of engineering. Their cost varies from $1,000 to $50,000. A model for evaluating software packages assists those selecting them:

Sanders, G. Larry et al. (1983). Model for the evaluation of computer software. *Computers & Industrial Engineering* 7:309–315.

As this book goes to press, a new abstract journal is being announced:

SAFE–Software Abstracts for Engineers. Quarterly. CITIS (Construction Industry Translation and Information Service), Dublin, New York.

23

License Review

This chapter will bypass the books and articles on the advantages of obtaining the PE and deal with examples of the literature that supply information on registering and preparing for the examinations which the engineer must pass to become a registered/licensed professional engineer (PE) or engineer-in-training (EIT).

Registration is usually achieved by compliance with three requirements: education, satisfactory experience and examination. The examination is given in two parts: the Fundamentals of Engineering Examination (FE) (leading to certification as Engineer-in-Training); and Principles and Practice of Engineering Examination (PE) (leading to licensure as a professional engineer). Many engineering students take the FE during their senior year; four years later, after qualifying experience, they are eligible to take the PE.

Registration requirements vary, and states and territories of the United States have their own engineering registration boards. A list, *State Engineering Registration Boards*, with their addresses and phone numbers, is available from:

National Society of Professional Engineers
2029 K Street, NW
Washington, D.C. 20006.

Nearly all states use the Fundamentals of Engineering and Principles and Practice examinations prepared by the National Council of Engineering Examiners (NCEE). The purpose of the Council, as stated in its constitution, is "to provide an organization through which State Boards may act and counsel together to better discharge their responsibilities in regulating the practice of engineering and land surveying as it relates to the welfare of the public in safeguarding life, health and property."

Information on the examinations and copies of the study guides prepared by the Council can be obtained by writing to:

National Council of Engineering Examiners
P.O. Box 1686
Clemson, South Carolina 29631.

The study guides are of two types: an examinations book and various questions pamphlets.

PROFESSIONAL ENGINEERING EXAMINATIONS BOOK

Professional Engineering Examinations, National Council of Engineering Examiners, Clemson, S.C.

Vol. 1. *1965–1971*. 1972, 412 pp. A compilation of questions from examinations administered by State boards from EIT and PE examinations during 1965–71 in chemical, civil, electrical and mechanical engineering.

Vol. 2. *Solutions: Professional Engineering Examinations (1965–71)*. 1972, 960 pp.

Vol. 3. *Professional Engineering Examinations (1972–76)*. 1977, 200 pp. Study guide for Fundamentals of Engineering and Principles and Practice of Engineering examinations. Contains sample Fundamentals of Engineering examination. Principles and Practice portion includes following disciplines: agricultural, ceramic, industrial, manufacturing, nuclear, petroleum, sanitary, structural and aeronautical/aerospace.

TYPICAL QUESTIONS PAMPHLETS

Typical Questions Pamphlets: Fundamentals of Engineering (1979), *Principles and Practice of Engineering* (1980), *Fundamentals and Principles and Practice of Land Surveying* (1980). National Council of Engineering Examiners, Clemson, S.C.

Outline topics covered in each examination, give overview of current examination format and examples of examination questions. Include aerospace/aeronautical, agricultural, ceramic, industrial, manufacturing, nuclear and petroleum engineering.

Sample Fundamentals of Engineering Examination. National Council of Engineering Examiners, Clemson, S.C., 1983, 100 pp.

Examples of other publications for use in preparing for the PE or EIT examination follow.

Engineer-in-Training License Review (9th ed.) Donald G. and Dean C. Newnan, Engineering Press, Inc., San Jose, Calif., 1981, 266 pp.

Material arranged according to 11 subjects in national EIT examination; typical problems with step-by-step solutions.

Engineer in Training Review Manual (6th ed.) Michael R. Lindeburg, Professional Publications, Inc., San Carlos, Calif., 1982, 760 pp.

Review of four-year engineering curriculum and reference for the E-I-T examination.

Engineering Economic Analysis (2nd ed.) Donald G. Newnan, Engineering Press, San Jose, Calif., 1983, 519 pp.

Includes topics contained in EIT and PE examinations: cost, depreciation, taxes, etc.

Engineering Fundamentals for Professional Engineers' Examinations (2nd ed.) Lloyd M. Polentz, McGraw-Hill Book Co., New York, 1980, 386 pp.

Fundamentals of Engineering Review. Iowa State University Research Fdn., Inc., Kendall/Hunt Publishing Co., Dubuque, Ia., 1981, 212 pp.

For engineering graduates in preparing for Engineer-in-Training examination. Accompanies series of videotapes prepared to help Iowa State's graduates.

Professional Engineer's License Guide: What You Need to Know and Do to Obtain PE (and EIT) Registration (4th ed.). Joseph D. Eckard, Herman Publishing, Inc., Boston, 1982, 125 pp.

Chapters on registration requirements and schedules; preparing and taking the examinations; addresses of

State Boards and of engineering organizations that produce the examinations or offer relevant information.

Civil Engineering License Review (8th ed.) Donald G. Newnan, Engineering Press, Inc., San Jose, Calif., 1980, 314 pp.

Problems from the seven categories of the Uniform Engineering Examination with step-by-step solutions.

Electrical Engineering License Review (5th ed.) Lincoln D. Jones and James A. Lima, Engineering Press, Inc., San Jose, Calif., 1980, 282 pp.

Electronics Engineering for Professional Engineers' Examinations. Charles R. Hafer, McGraw-Hill Book Co., New York, 1980, 309 pp.

Mechanical Engineering License Review (3rd ed.) Richard K. Pefley and Donald G. Newnan, Engineering Press, Inc., San Jose, Calif., 1980, 376 pp.

Study Guide to the Professional Engineers' Examination for Industrial Engineers. R.P. David et al., Institute of Industrial Engineers, Norcross, Ga., 1983, 500 pp.

There are numerous other books describing the registration process and supplying review material. The National Society of Professional Engineers' *Selected Bibliography on Professional Engineers and Fundamentals (Engineering-Training) Examination Preparation* (NSPE Publication no. 2201, revised October 1981) includes several of the books listed above and many others. These and still others can be located in the subject section of library card catalogs by searching under "Engineering–Examinations, questions, etc." or in *Books in Print* or *Scientific and Technical Books and Serials in Print* under the heading, "Engineering–Examinations, questions, etc." Material on registration appears frequently in journal articles. Two sources of such articles are the index and abstract services, *Applied Science and Technology Index* and *Engineering Index*, where articles can be located by searching under Engineers–Registration.

The National Council of Engineering Examiners publishes *NCEE Registration Bulletin* in February, April, June, October and December. The *Bulletin* prints news of member Boards, revision of test specifications and other current information on Council activities.

24

Preparation of Technical Reports

Engineers must communicate in a variety of forms–resumés to secure initial and subsequent positions; memos and letters to administrators, other engineers and associates, vendors; papers presented at meetings and conferences. Internal memos, letters, travel reports and laboratory reports constitute a body of engineering information germane to every company or other organization, with every engineer being responsible for writing them. Colleagues and supervisors will introduce them to this bank of information, but they must be prepared to contribute to it in an effective manner, to report and to convince.

Probably, the form of communication engineers use most frequently is the technical report. The technical report serves as the earliest transmittal of research and development results. If the research is being conducted under a government contract, a technical report is obligatory, and its form and frequency are usually spelled out in the contract. For an internal project, the report is usually required for the company's or institution's files. All research results are not disseminated, because they sometimes contain proprietary or classified information, but when they can be reported, sharing research results prevents the duplication of research and makes the knowledge gained from projects available to others who can use it in their research.

Preparing the technical report is part of the engineer's performance of research and development projects. Technical reports, beyond being prompt, must be clear and concise. Learning the skills needed to write such reports should be part of every engineer's education or acquired during the early years of a professional career. Engineers should be impressed with their importance in communicating their ideas and the results of their work to others in the organization in which they are employed and to the scientific and technical community. In many projects, writing of the report is the only "product." There is a saying, "The work isn't done until the report has been written." To be sure the report is read as well, authors must consider a quotation in an article by Ben Weil, a technical writer, that sums up the raison d'etre of reports: "So tell me quick and tell me true, or else I haven't time for you. Not how your study came to be but what you found that can help me."*

This chapter does not deal directly with technical writing, but with books and journals that will help the engineer to master the preparation of technical reports. One chapter cannot accomplish a full discussion of such preparation; this one will cover materials on the elements of the report and its preparation and refer readers to books and journal articles on technical writing. Results of a survey in 1978 of 348 engineers listed in *Engineers of Distinction* found that the respondents

---

*Weil, Ben H. (1981). Presenting R&D clearly. *ChemTech.* 11:720–723.

spend almost 25% of their time in technical writing and an additional 31% in working with materials that others have written. The author writes that these engineers attribute their advancement largely to their ability to communicate well and deplore the poor quality of much of the material they must read.*

Engineering colleges may eventually require courses in technical writing. Meanwhile, engineering students and professional engineers can benefit from the books and journal articles published by other engineers and technical writers. The large number published indicates the necessity for practical information on this subject. The monthly issues of *Applied Science and Technology Index*, for example, invariably include articles on technical writing. They usually appear under the headings, "Engineering Reports," "Technical Reports," "Technical Writing." An institute of Electrical and Electronic Engineers journal, *IEEE Transactions on Professional Communication*, is devoted to the numerous forms of communications used by the scientific/technical community. In the course of a recent year, this journal covered job instructions, talking in public, policies and procedures manuals, graphics, resumes and interviews.

To find books on writing the technical report, the engineer should consult library card catalogs under the subject headings "Technical writing" or "Technical reports." To find what books are in print and therefore available for purchase, engineers can use *Scientific and Technical Books and Serials in Print*, checking the headings, "Technical writing," "Technical reports," "Technical illustration," "Communication of technical information." Most books on technical report writing contain sections on the various elements of the report: abstract, annotations, references, bibliographies, tables, illustrations. Such chapters will be found in the following examples.

---

*Davis, Richard M. (1978). How important is technical writing—a survey of the opinions of successful engineers. *J. Technical Writing and Communication*. 8:207-216.

Scientists Must Write: A Guide to Better Writing for Scientists, Engineers and Students. Robert Barrass, Chapman & Hall, London, 1978, 176 pp.

> Describes the importance of writing to the groups named in the title, of the planning, organizing and communicating of their research results. Section on reports offers guidelines on elements such as tables of contents, lists of references, summary, etc.

How to Write and Publish a Scientific Paper. Robert A. Day, ISI Press, Philadelphia, 1979, 160 pp.

> The author offers basic principles for preparing a readable paper. Chapters on preparing abstracts, tables and illustrations. Two chapters contain material on working with editors and printers.

Effective Technical Communication. Ann Eisenberg, McGraw-Hill, New York, 1982, 355 pp.

> Overview including material on style, basic guidelines, types. Includes chapters on writing proposals and letters and on preparing speeches, as well as reports. Chapter on abstracts is comprehensive.

How to Write and Publish Engineering Papers and Reports. H.B. Michaelson, ISI Press, Philadelphia, 1982, 158 pp.

> Emphasizes the writing strategies for customizing reports, papers, etc. to the information needs and interests of the audiences to whom they are directed.

Principles of Communication for Science and Technology. Leslie A. Olsen, McGraw-Hill Book Co., New York, 1983, 414 pp.

Technical Reports Standards: How to Prepare and Write Technical Reports. Lawrence R. Harvill and Thomas L. Kraft, Banner Books International, Sherman Oaks, Calif., 1977, 54 pp.

Creating the Technical Report. Steven Schmidt, Prentice-Hall, Englewood Cliffs, N.J., 1983, 160 pp.

The Writing System for Engineers and Scientists. Edmond H. Weiss, Prentice-Hall, Englewood Cliffs, N.J., 1982, 274 pp.

> Includes analysis of selection of purpose, objectives and audience. Chapters on outlining, proposal writing and instruction manuals.

Beyond assembling the elements of the report, the engineer must develop a readable style. The engineer must consider the audience and assure that they will understand what has been written. Writing "tips" appear frequently in a broad spectrum of the scientific and technical journals.

Some companies or institutions employ technical writers and editors whose responsibility it is to write or rewrite technical reports and other communications, but the majority do not have secretaries or other persons with the time or talent to take over this responsibility; thus, it remains the engineer's. The books and journal articles cited in this chapter were chosen from the many designed to develop or improve technical writing skills and technical reports preparation.

Often, books or journal articles treat the elements of a report exclusively. The following book on abstracting is an example:

Art of Abstracting. Edward T. Cremmins, ISI Press, Philadelphia, 1982, 150 pp.

After a report is written, the data in it may be expanded or edited and presented at a symposium or conference or submitted to a journal. It may ultimately be incorporated into a book, with additional background material, illustrations, references, etc. The publishers of these forms usually supply the most explicit help. Generally labelling them "Instructions to the

Author," they set up outlines for preparation of material, i.e., length of abstract, form for presenting data, references, diagrams, etc. An example of such instructions is "Information for Authors" which appears on the back cover of most issues of the *IEEE Transactions.*

Style manuals offer guidance to writers in any field or format. Two such manuals are:

Technical Writing Style Guide. U.S. Nuclear Regulatory Commission, NUREG-0650, 1980.

Elements of Style (2nd Ed.). William Strunk, Jr. and E.B. White, Macmillan, New York, 1972, 78 pp.

The books and articles cited in this chapter serve only as examples of the assistance to engineers to be found in the literature of technical writing. Further information, as noted above, can be obtained in libraries.

25

Libraries, Information Centers and Information Brokers

If an engineer works for a company or agency that has a library or information center, he has not only information resources immediately available, but also the assistance of the librarian or information specialist in obtaining material from or access to the collections of other libraries or information centers.

These internal information systems vary widely, however, in their collections, staffs, and services. They can range from a bookcase presided over by a clerical assistant to very sophisticated operations run by dozens of information specialists, such as those at Exxon or Bell Telephone Laboratories. Many will have files of internal documents, such as laboratory notebooks, project memos, trip reports, and proprietary documents, that are not available through the usual library channels.

For further information on company information practices, see the book *Industrial Information Systems* cited at the end of the chapter.

Unfortunately, the majority of engineers are employed by companies without such facilities. They must know for themselves what information resources are available and how they can retrieve them. The purpose of this book is to inform them of the various formats. In some of the chapters, information is provided on how to obtain copies of the specific materials discussed, e.g. Chapter 12 on "Patents" and Chapter 11 "Standards and Specifications." The objective of this chapter is to add to this specific information and list some of the libraries and other resources that engineers can tap, or where they can go to do research, borrow materials or obtain copies of articles.

LIBRARIES

Even if an engineer has no company librarian to turn to, he should be aware that there are usually nearby public and academic libraries which can help locate the information he needs.

Public libraries usually impose only residential restrictions. Any person living within the geographic area served can use the collections, and is entitled to ask that materials not in the library be obtained on "inter-library loan" from another library. There is not usually a charge for inter-library loan requests, but they normally take several weeks to be filled.

Most state and private universities and colleges permit the use of their libraries on the premises. Some even offer "courtesy" borrower cards to engineers upon presentation of a plant pass or business card. Others have "Friends" associations whose paid membership includes borrowing privileges, and a few even have fee-based information services similar to the information brokers to be discussed later.

There are four major libraries in the United States with exceptionally large collections in science and engineering that engineers should know about.

Each of these are briefly described below.

John Crerar Library

35 W. 33rd St.
Chicago, Ill. 60616
312-225-2526

Subjects: Sciences–physical, biological, medical; Engineering

Holdings: 700,000 volumes; 7000 subscriptions

Services: Interlibrary loans; copying, contract bibliographic services; library open to public.

Engineering Societies Library

United Engineering Center
345 E. 47th St.
New York, N.Y. 10017
212-644-7611

Subjects: Engineering

Holdings: 250,000 volumes; 5500 subscriptions

Services: Copying and literature searches available as paid services; library open to public.

Linda Hall Library

5109 Cherry St.
Kansas City, Mo. 64110
816-363-4600

Subjects: Sciences–physical, life; Engineering

Holdings: 525,000 volumes; 15,600 subscriptions

Services: Interlibrary loans; copying; library open to public.

New York Public Library, Science & Technology Research Center

5th Ave. & 42nd St.
New York, N.Y. 10018
212-930-0573

Subjects: Physical sciences; Engineering

Holdings: 900,000 volumes; 4900 subscriptions

Services: Reference service by phone and letter; center open to public for reference use only.

INFORMATION CENTERS

The difference between a library and an information center is not sharp and distinct. Information centers tend to differ from libraries in the nontraditional materials in their collections, and in the specialized services they offer, but the choice of name is sometimes decided by how a manager feels about the traditional implications of the word "library." A special library today can be declared an information center tomorrow with no discernible change in activities.

There are a substantial number of agencies called information centers that could be very helpful to engineers. These centers offer a variety of special services, including: generation and maintenance of handbooks and data books; state-of-the-art studies; critical reviews and technology assessments; bibliographic inquiries; and current awareness, marketing, and promotion publications. The essential information for these centers is given below; each can supply brochures and other materials describing their services.

Chemical Kinetics Data Center

National Bureau of Standards
Bldg. 222, Rm. A166

Washington, DC 20234
301-921-2174

Chemical Propulsion Information Agency (CPIA)

Johns Hopkins University
Applied Physics Lab.
Laurel, MD 20707
301-953-7100, Ext. 3154

Coastal Engineering Information Analysis Center

Fort Belvoir, VA 22060
202-235-7386

Cold Regions Science and Technology Information Analysis Center

Hanover, NH 03755
603-643-3200, Ext. 339

Concrete Technology Information Analysis Center

Army Engineer Waterways Experiment Station
Vicksburg, MS 39180
601-634-3264

Conservation & Energy Inquiry & Referral Service

Dept. of Energy
Box 890
Silver Spring, MD 20907
800-523-2929; 800-462-4983 (PA)

Data and Analysis Center for Software

Rome Air Development Center
Griffiss AFB, NY 13441
315-336-0937

DOD Nuclear Information and Analysis Center (DASIAC)

Kaman-Tempo
Santa Barbara, CA 93102
805-963-6400

Earthquake Engineering Research Center

National Information Service for Earthquake Engineering
University of California at Berkeley
47 St. & Hoffman Blvd.
Richmond, CA 94804
415-231-9403

Engineering Physics Information Centers (EPIC)

Oak Ridge National Laboratory
Box X
Oak Ridge, TN 37830
615-574-6176

Environmental Science Information Center

National Oceanic & Atmospheric Administration
Environmental Data & Information Service
11400 Rockville Pike, Code D8
Rockville, MD 20852
301-443-8137

Guidance and Control Information Analysis Center (GACIAC)

Illinois Institute of Technology
Research Institute (IITRI)
Chicago, IL 60616
312-567-4519

Hydraulic Engineering Information Analysis Center

Army Engineer Waterways Experiment Station
Vicksburg, MS 39180
601-634-2795

Infrared Information and Analysis Center (IRIA)

Environmental Research Institute of Michigan
Ann Arbor, MI 48107
313-994-1200, Ext. 214

Institute of Polar Studies

Ohio State University
125 S Oval Mall
Columbus, OH 43210
614-422-6531, 6532

International Atomic Energy Agency (IAEA), Nuclear Data Section

Box 100
A-1400 Vienna, Austria
0222-2360-1709

Machinability Data Center

Metcut Research Associates Inc.
3980 Rosslyn Drive
Cincinnati, OH 45209
513-271-9510

Metal Matrix Composites Information Analysis Center (MMCIAC)

Kaman-Tempo
Santa Barbara, CA 93102
805-963-6497

Metals and Ceramics Information Center (MCIC)

Battelle-Columbus Laboratories
Columbus, OH 43201
614–424–5000

National Oceanographic Data Center (NODC)

Environmental Data & Information Service, NOAA
Page Bldg. No. 1
2001 Wisconsin Ave., NW
Washington, DC 20235
202–634–7500

Nondestructive Testing Information Analysis Center (NTIAC)

Southwest Research Institute
San Antonio, TX 78284
512–684–5111, Ext. 2362

Nuclear Safety Information Center

Box Y
Oak Ridge, TN 37830
615–574–0391

Pavements and Soil Trafficability Information Analysis Center

Army Engineer Waterways Experiment Station
Vicksburg, MS 39180
601-634-2795

Plastics Technical Evaluation Center

Army Armament Research & Development Command
Dover, NJ 07801
201–328–4222

Reliability Analysis Center (RAC)

Rome Air Development Center
Griffiss AFB, NY 13441
315-330-4151

Rubber & Plastics Research Association (RAPRA)

Shawbury, Shrewsbury SY4 4NR
United Kingdom
0930-250383

Shock and Vibration Information Center

Naval Research Laboratory
Washington, DC 20375
202-767-2220

Shock Wave Data Center

Lawrence Livermore National Laboratory
Box 808
East Ave.
Livermore, CA 94550
415-422-7216

Soil Mechanics Information Analysis Center

Army Engineer Waterways Experiment Station
Vicksburg, MS 39180
601-634-3475

Tactical Technology Center

Battelle-Columbus Laboratories
Columbus, OH 43201
614-424-7010

Thermodynamics Research Center

Texas A & M University
College Station, TX 77843
713-845-4940

Thermophysical and Electronic Properties Information Analysis Center (TEPIAC)

Purdue University
West Lafayette, IN 47906
317-494-6300

Water Research Centre

Information Service on Toxicity & Biodegradability
Elder Way, Stevenage SGL 1TH
United Kingdom
0438-2444

INFORMATION BROKERS AND ONLINE ORDERING

Information brokers are a relatively recent occurrence in the library world. Many of these companies were started by librarians in a tight job market going into business for themselves, offering document delivery, literature searches, and other information services. The advantages brokers offered were obvious—speedier delivery than interlibrary loan, and the need to pay for services only when they were used. For small companies that cannot afford a full-time librarian, the information brokers fill the bill very well. Many special libraries have now come to depend on them also to obtain materials not owned in the information center.

In a related area, some of the major computer-search services now are offering online ordering of documents. Dialog's DIALORDER and SDC's ORBDOC online ordering services will transfer requests to a broker or supplier of the user's choice.

There are several publications that provide listings of information brokers, along with a variety of other services and products. The reader should also try the yellow pages of the telephone directory under "Information Services & Bureaus" and "Library Research Service" for nearby information suppliers.

Directory of Special Libraries and Information Centers. (8th ed.). Brigitte T. Darnay (ed.), Gale Research Co., Detroit, 1983, (3 volumes in 4 parts)

> "A guide to special libraries, research libraries, information centers, archives, and data centers maintained by government agencies, business, industry, newspapers, educational institutions, nonprofit organizations, and societies in the fields of science and technology, medicine, law, art, religion, the social sciences, and humanistic studies."

Information Industry Market Place, 1983. R. R. Bowker, New York, 1982, 282 pp.

> An international directory of information products and services. Includes listings of machine-readable databases, print products, online vendors, library networks, telecommunication networks, information collection and analysis centers, information brokers, terminal manufacturers, consultants and support services, associations, online user groups, government and international agencies, conferences and courses, reference books, periodicals and newsletters.

Encyclopedia of Information Systems and Services. (5th ed.). John Schmittroth (ed.), Gale Research Co., Detroit, 1982, 1242 pp.

> "An international guide to computer-readable data bases,

data base producers and publishers, online vendors and time-sharing companies, telecommunications networks, videotex/teletext systems, information retrieval software, library and information networks, bibliographic utilities, library management systems, fee-based information services, data collection and analysis centers, community information and referral systems, and related consultants, service companies, associations, and research centers."

Directory of Fee-based Information Services, 1982. Kelly Warnken (ed.), Information Alternative, Chicago, 1982, 114 pp.

Annual listing of information brokers, freelance librarians, independent information specialists, library consultants, information retailers, information packagers, public and academic libraries and others providing library and information services for a fee.

REFERENCES

Industrial Information Systems: A Manual for Higher Managements and Their Information Officer/Librarian Associates. Eugene B. Jackson and Ruth L. Jackson, Hutchinson Ross Publishing Co., Stroudsburg, Pa., 1979, 336 pp.

Appendix

PUBLISHERS' ADDRESSES

This list of publishers of scientific and technical books is highly selective, providing only the names and addresses of those estimated to be publishing the greatest number of titles and those whose addresses are most frequently requested in a typical engineering library. Societies, as well, qualify for listing. More comprehensive lists are available in *Books in Print*, or one of the international or specialist publishers' directories.

Commercial Publishers

United States

Academic Press, Inc.
Orlando, Fla. 32887
800-321-5068

Addison-Wesley Publishing Co., Inc.
Jacob Way
Reading, Mass. 01867
617-944-3700

Ann Arbor Science Publishers
c/o Butterworth Publishers, Inc.
10 Tower Office Park
Woburn, Mass. 01801
617-935-9361

CRC Press
2000 Corporate Blvd.
Boca Raton, Fla. 33431
305-994-0555

Marcel Dekker, Inc.
270 Madison Ave.
New York, N.Y. 10016
212-696-9000

Elsevier Science Publishing Co., Inc.
52 Vanderbilt Ave.
New York, N.Y. 10017
212-867-9040

Engineering Press, Inc.
P.O. Box 1
San Jose, Calif. 95103
408-258-4503

Gordon & Breach Science Publishers, Inc.
1 Park Avenue
New York, NY 10016
212-689-0360

Gulf Publishing Co.
Box 2608
Houston, TX 77001
713-529-4301

Industrial Press, Inc.
200 Madison Ave.
New York, N.Y. 10157
212-889-6330

McGraw-Hill Book Co.
1221 Avenue of the Americas
New York, N.Y. 10020
212-997-1221

MIT Press
28 Carleton St.
Cambridge, Mass. 02142
617-253-2884

Pergamon Press, Inc.
Maxwell House
Fairview Park
Elmsford, N.Y. 10523
914-592-7700

Plenum Publishing Corp.
233 Spring St.
New York, N.Y. 10013
212-620-8000

Prentice-Hall, Inc.
Rte. 9w
Englewood Cliffs, N.J. 07632
201-592-2000

Reston Publishing Co., Inc.
11480 Sunset Hills Rd.
Reston, Va. 22090
703-437-8900

W. B. Saunders Co.
Washington Square
Philadelphia, Pa. 19105
215-574-4700

Charles C. Thomas, Publishers
2600 S. First St.
Springfield, Ill. 62717
217-789-8980

Van Nostrand Reinhold Co.
135 W. 50th St.
New York, N.Y. 10020
212-265-8700

John Wiley & Sons, Inc.
605 Third Ave.
New York, N.Y. 10158
212-850-6418

Great Britain (U.S. addresses given when available)

Allen & Unwin, Inc.
9 Winchester Terrace
Winchester, Mass. 01890
617-729-0830

Edward Arnold
300 North Charles St.
Baltimore, Md. 21201

Butterworths Publishers, Inc.
10 Office Tower Park
Woburn, Mass. 01801
617-933-8260

Chapman & Hall Ltd.
11 New Fetter Lane
London EC4 P4EE

Elsevier Applied Science Publishers
[Books can be ordered through Elsevier. See above]
22 Rippleside Commercial Estate
Ripple Rd.
Barking, Esses 1G11 OSA
England

Oliver & Boyd
Robert Stevenson House
1-3 Baxter's Place
Leith Walk
Edinburgh
Scotland EH1 3BB

Oxford University Press, Inc.
200 Madison Ave.
New York, N.Y. 10016
212-679-9300

Taylor & Francis Ltd.
4 John St.
London WC1 N2ET

Germany

Akademie-Verlag
Leipziger Str 3–4
DDR–1080 Berlin

Walter de Gruyter
200 Saw Mill River Rd.
Hawthorne, N.Y. 10532
914–747–0110

Springer-Verlag
175 Fifth Ave.
New York, N.Y. 10010
212–460–1584

France

Dunod, Librairie
17, Rue Remy-Dumoncet
F–75686 Paris

Hermann, Editeurs des Sciences et des Arts
293, Rue Lecourbe
F–75015 Paris

Masson S. A., Editeur
120 Blvd. Saint-Germain
F–75280 Paris

Netherlands

North-Holland Publishing Company
P.O. Box 211

1000 AE Amsterdam
The Netherlands
[in U.S. and Canada, Elsevier (see above)]

Societies

American Institute of Aeronautics and Astronautics
1633 Broadway
New York, N.Y. 10019
212-581-4300

American Nuclear Society
555 N. Kensington Ave.
LaGrange Park, Ill. 60525
321-352-6611

American Society of Civil Engineers
345 E. 47th St.
New York, N.Y. 10017
212-705-7518

American Society of Heating, Refrigerating and Air Conditioning Engineers
1791 Tullie Circle NE
Atlanta, Ga. 30329
404-636-8400

American Society for Metals
9275 Kinsman Rd.
Metals Park, Ohio 44073
216-338-5151

American Society of Mechanical Engineers
345 E. 47th St.
New York, N.Y. 10017
212-705-7722

Association for Computing Machinery
1133 Ave. of the Americas
New York, N.Y. 10036
212-265-6300

Institute of Electrical & Electronics Engineers
(orders) 445 Hoes Lane
Piscataway, N.J. 08854
201-981-0060

Institute of Industrial Engineers
25 Technology Park
Atlanta, Norcross, Ga. 30092
404-449-0460

Institution of Electrical Engineers
(Books, etc. may be ordered from IEEE)

Institution of Civil Engineers
107 Great George St.
London SWIP 3AA

Institution of Mechanical Engineers
Mechanical Engineering Publications Ltd.
P.O. Box 24
Northgate Ave.
Bury St.
Suffolk 1P32 6BW
England

REFERENCES

Book Publishers Directory: A Guide to New and Established, Private and Special Interest, Avant-garde and Alternative, Organization and Association, Government and Institution Presses

(2nd ed.). Annie M. Brewer and Elizabeth A. Geiser (eds.), Gale Research Co., Detroit, 1979, 668 pp.

Subject, geographic indexes.

Books in Print. R. R. Bowker Co., New York. Annual. Published since 1948.

List of publishers at end of title entries.

International Academic and Specialist Publishers Directory. Tim Clarke, R. R. Bowker, New York, 1975, 555 pp.

Arranged by country, publisher, subject indexes.

Index

Abstracting, 181
Abstracts, printed and computerized, 8–18
Addresses
  local representatives of companies, 127
  manufacturers, 127
  publishers, 195-203
Almanacs, 121
American Association of Engineering Societies, 149–150
American Men and Women of Science, 75
American National Standards Institute, 89–90
American Society for Testing and Materials, 85-87
Annual reports, 120
Annual reviews . . ., 117
ANSI (*see* American National Standards Institute)
Applied Science & Technology Index, 9, 14
Associations (*see* Professional societies)

ASTM (*see* American Society for Testing and Materials)
Atlases, 153–156
Atomindex, INIS, 34
Audio-visual materials, 161–163
AV (*see* Audio-visual materials)

Back issues of periodicals (*see* Periodicals, back issues)
Biographical directories, 75–76
  definition, 74
Books, 24–29
  foreign, 27
  in print, 26
  ordering, 26–27
  out-of-print, 27
  reviews, book, 25
Brokers (*see* Information brokers)
Buyers' guides, 119, 121, 126

Catalogs, manufacturers (*see* Catalogs, trade)
Catalogs, trade, 125–126
  ASK IV, 128
  collections, 127–128
  Sweet's files, 127
  Thomcat, 127
Charts, 153
Chemical Engineers' Handbook, 40
Civil Engineers, Standard Handbook for, 41
Codes (*see* Standards)
Company catalogs (*see* Catalogs, trade)
Company magazines (*see* House organs)
Composite Index for CRC Handbooks, 51
Computer science
  abstracts, 15
  encyclopedias, 64
Computer-search databases (*see* Databases)
Computer searching, 11–13 (*see also* Databases)
Conference Papers Index, 144
Conferences, 140–145
  forthcoming, 141–142
  publications of, 142–145
Consultants and Consulting Organizations Directory, 78
Cover-to-cover translations (periodicals) (*see* Periodicals, translated)
CRC Handbook of Tables for Applied Engineering Science, 55
Crerar, John, Library (*see* John Crerar Library)

Critical data, 53
Current Contents, 9-10, 15

Data books (*see* Handbooks)
Databases 14-23 (*see also* Computer searching)
bibliographic, 14-18
conference proceedings, 144-145
dissertations and theses, 123-124
patents, 106-109
standards and specifications, 92-95
technical reports, 33-36
Databases, full text, 23
Databases, non-bibliographic, 19-23
Databases, numeric, 19-20
Databases, referral, 19, 20-22
Defense Department specifications
index, 91
Defense Technical Information Center
TAB Index, 35-36
Definition dictionaries, 67-70
Depositories, Patent (*see* Patent depositories)
Derwent Publications, Ltd., 103-104, 109-110
Dictionaries, 66-72
Directories, 73-82
Directory of Published Proceedings, 143
Dissertation Abstracts International, 123-124
Dissertations, 122-124

E-I-T (*see* Engineer-in-training)
Electrical and Computer Engineering, Fundamentals Handbook of, 40
Electronic Yellow Pages, 80-82
Electronics Engineers' Handbook, 40
Encyclopedia of Associations, 76-77, 147-148
Encyclopedia of Chemical Technology, 63-64
Kirk-Othmer/Online, 21
Encyclopedias, 60-65
Energy Research Abstracts, 33
Engineer-in-training (E-I-T) examinations, 171-176
Engineering Index, 9-10, 16
Engineering meetings (*see* Conferences)
Engineering reports (*see* Technical reports)
Engineering societies (*see* Professional societies)

Engineering Societies Library, 26, 185
Engineering standards, 83–85 (*see also* Standards)
Engineers' handbooks (*see also* Handbooks
  Selected examples . . ., 41
Examinations–Professional engineers, 171–176

Federal specifications, 93 (*see also* Standards)
  databases, 93–94
  index, 91
Films
  directories, 162
Foreign-language dictionaries, 70–72
Formulas, 44
Fundamentals of Engineering Examinations (FE), 171–176
  bibliography, 175

Gazetteers, 154–155
Geological Survey (*see* United States Geological Survey)
Glossaries (*see* Dictionaries)
Government Reports Announcements & Index, 33

Handbook of Chemistry and Physics, 55
Handbooks, 38–52
  bibliographies, 55
Hi Tech Patents, 105, 107
House organs, 129

IEC (*see* International Electrotechnical Commission)
IHS (*see* International Handling Services)
Index to Scientific and Technical Proceedings, 144
Indexes, periodical
  annual, 5
  cumulative, 5
Indexes, printed and computerized, 8–18
Industrial Engineering, Handbook of, 40
Industrial Research Laboratories of the U.S., 77
Information brokers, 192–194
Information centers, 186–192
Information Marketing International, 98–99
INIS Atomindex, 34
INSPEC, 15–17
Interlibrary loan, 4, 184
International Critical Tables, 53

International Electrotechnical Commission, 89-90
International Handling Services, 97
databases, 82, 92-95
printed index, 90-91
International Organization for Standardization, 89-90
ISO (*see* International Organization for Standardization)

John Crerar Library, 135, 185
Journals (*see* Periodicals)

Kirk-Othmer (*see* Encyclopedia of Chemical Technology)

Libraries, 183-194
License reviews, 171-176
Linda Hall Library, 185

Machine searching (*see* Computer searching)
MacRAE's Blue Book, 79
Magazines (*see* Periodicals)
Manuals (*see* Handbooks)
Manufacturers
directories, 127
Maps, 151-156
bibliographies, 154
dealers, 153
map collections, 153, 155-156
Masters Abstracts, 124
Mathematical tables (*see* Tables, mathematical)
McGraw-Hill Encyclopedia of Science & Technology, 62
yearbook, 119
Mechanical Engineers' Handbook, 41
Meetings (*see* Conferences)
Military specifications, 93 (*see also* Standards)
databases, 93-94
index, 91

National Aeronautics & Space Administration
STAR index, 35
National Bureau of Standards, 54, 87-88
National Center for Standards and Certification Information, 95-96
National Council of Engineering Examiners, 172-173, 176

National Society of Professional Engineers, 172, 175
National Standards Association, 98
databases, 93-94
National Technical Information Service, 32, 36, 105
National Translations Center, 135
New York Public Library, 186
New Technical Books, 28
NTIS (*see* National Technical Information Service)
Nuclear Science Abstracts, 34

Online searching (*see* Computer searching)
Ordering
books, 26–27
maps, 153
patents, 109–114
periodicals, 3
standards, 96–99
technical reports, 36

Patent depositories, 110–114
Patent Office, U.S. (*see* United States Patent and Trademark Office)
Patents, 100–115
databases, 106–109
depositories, 110–114
obtaining copies, 109–114
printed indexes, 101–105
PE (*see* Professional engineer)
Periodicals, 1–6
back issues, 5
directories, 3
holdings of libraries, 4
indexes, 5
obtaining copies of articles, 4
special issues, 5, 119, 121, 126–127
subscriptions, 3
translated, cover-to-cover, 5
union lists, 4
Periodicals, abstracting and indexing (*see* Abstracts, printed and computerized)
Principles and Practice of Engineering Examination (PE), 171–176
bibliography, 175
Proceedings (*see* Conferences)
Proceedings in Print, 143
Product literature, 125–126
Professional engineer (PE) examinations, 171–176

Professional societies, 146–150
as publishers, 25
directories, 119
Programs, computer, 164–166
Properties, physical, etc., tables of, 55–58
Publishers, 24-25, 195-203
Publishers' addresses, 195-203
Purchasing (*see* Ordering)

Registration Boards, State Engineering, 172
Reports (*see* Technical reports)
Research Centers Directory, 77
Reviews, 116–118
book reviews, 25
Index to Scientific Reviews, 118
indexes, 118
journal articles (review articles), 118

Science Abstracts, 15, 17
Science Citation Index, 18
Science Indicators, 159
Searching services (*see* Computer searching)
Selected Water Resources Abstracts, 35
Societies (*see* Professional societies)
Software, 164–170
directories, 167–168
evaluation, 170
exchanges, 167, 169
information and service centers, 168–170
periodicals, 166
Special issues of periodicals, 5, 119, 121, 126–127
Specifications (*see* Federal specifications, Military specifications)
Standards, 83–99
databases, 92–95
definition of, 83
formulators, 85–90
indexes, 90–92
obtaining copies of, 96–99
Tech-net, 128
State-of-the-art surveys, 118
Statistical Abstract of the United States, 160
Statistical Sources, 158
Statistics, 157–160
directories, 157–160
in periodicals, 159
Style guides, 182
Subscriptions, periodicals (*see* Periodicals, subscriptions)
Surveys (*see* Reviews)

Sweet's Catalog files, 79-80, 127
Symposia (*see* Conferences)

Tables
bibliographies, 55
in handbooks, 40–52
mathematical, 55–57, 59
of properties, 55–58
statistical (*see* Statistics)
Tech-net, 82, 94–95, 128
Technical information (product), 125
Technical reports, 30–37
databases, 33–36
definition of, 30
indexes, 33–36
obtaining copies, 36
writing, 177–182
Technical writing (*see* Technical reports, writing)
Term dictionaries, 67–70
Thermophysical Properties of Matter, 54
Theses, 122–124
Thomas Register of American Manufacturers, 80, 127
Topographic maps (*see* Maps)
Trade literature, 125–130
Trade names, 128–130
Translations, 131–139
cover-to-cover, periodicals, 5
indexes, 134–136
National Translations Center, 135
Translators, 136–137

Ulrich's International Periodicals Directory, 3, 10, 129
Union lists (*see* Periodicals, union lists)
United States Geological Survey, 151–153, 155
United States Patent and Trademark Office, 102, 108, 110
patent depositories, 110–114
University Microfilms International, 122–124

Videotapes, 161

Water Resources Abstracts, Selected, 35
Who's Who directories, 75-76
Who's Who in Engineering, 76, 150
Who's Who in Technology Today, 76

Writing, technical (*see* Technical reports, writing)

Yearbooks, 119–121
(Yearbooks)
encyclopedias, 119
location in library card catalogs, 121
societies, 119

About the Authors

MARGARET T. SCHENK is Head of Collection Development at the Science and Engineering Library, State University of New York at Buffalo. Her career in scientific and technical libraries spans 35 years, and includes service for American Optical Company, Bell Aerospace Textron, and Cornell Aeronautical Laboratory. Ms. Schenk has served on various committees and as President of the Upstate New York Chapter of the Special Libraries Association, and is a recipient of the Chapter's award for outstanding service. Additionally, she is a member of the American Society for Engineering Education, and the New York Library Association. With coauthor James K. Webster, Ms. Schenk teaches a course on engineering information at SUNY at Buffalo. She received the B.A. and B.L.S. degrees from the University of Buffalo.

JAMES K. WEBSTER is Head of Reference at the Science and Engineering Library, State University of New York at Buffalo, where he served as Director from 1976–82. He is the author of *The Bibliographic Utilities: A Guide for the Special Librarian*, and he has written several articles for Special Libraries and other publications. Mr. Webster is a member of the Special Libraries Association where he has held a number of national and local offices, as well as the New York Library Association, American Society for Engineering Education, and Association of College and Research Libraries. He received the M.L.S. degree (1966) from SUNY College of Geneseo.